THE GREYWACKE

THE GREYWACKE

How a Priest, a Soldier and a School Teacher Uncovered 300 Million Years of History

NICK DAVIDSON

P
PROFILE BOOKS

This paperback edition published in 2022

First published in Great Britain in 2021 by
Profile Books Ltd
29 Cloth Fair
London
EC1A 7JQ
www.profilebooks.co.uk

1 3 5 7 9 10 8 6 4 2

Printed and bound in Great Britain by
CPI Group (UK) Ltd, Croydon, CR0 4YY

A CIP catalogue record for this book is available from the British Library.

ISBN 978 1 78816 378 1
eISBN 978 1 78283 626 1

For Leo, Felix, Emma and Zander who will one day, I hope, come to enjoy the rocks as much as I do

Contents

INTRODUCTION

Some years ago relatives of mine moved to the Welsh borders, and I began to spend long, exhilarating days walking the heather- and bracken-clad uplands of the Berwyn Hills. I liked to gaze down from Thomas Telford's nineteenth-century trunk road, now the A5, at a point known as Pont Glyn-diffwys. There, the land plunges hundreds of feet to the fast-flowing River Ceirw, and human life is tucked away in narrow lanes and hamlets. I tramped across moorland as the bracken turned to autumn gold and grouse took to the air, squawking in alarm. And I followed damp sheep tracks up steep river gullies overhung by ferns and moss.

Such a landscape seizes the imagination: it's impossible not to wonder how it came to exist. But as I started to read about the region's past, one place in particular kept drawing me back. On the western flank of the Berwyns, where the land drops gently towards the wind-flecked waters of Lake Bala, there are a series of moss-draped pits and quarries. Once upon a time, blocks of dull grey limestone were dragged out of the earth here to build nearby farms and outbuildings. The spot is known as the Gelli-grin, an old Welsh term meaning 'parched or withered woodland'. It isn't much to look at. Nevertheless, I learned it had played a remarkable role in the effort to unravel the history of our planet.

When nineteenth-century mining engineers and naturalists first started to classify the limestones, chalks and clays that cover much of England, the Welsh hills – along with much of upland Britain – remained a complete mystery. They were widely acknowledged to be old, perhaps among the oldest rocks in Britain, and were therefore presumed to hold the key to such questions as the age and origins of the Earth, and when and how life had begun: controversial matters in Victorian Britain. And yet there was a maddening problem. The rocks themselves were so twisted and apparently chaotic, the strata so difficult to trace, the fossil record so slender and obscure that they defied all attempts to place them in any order. As a result, a jumble of very different minerals was lumped together in a holdall category known as the Greywacke, an anglicisation of the German mining term *Grauwacke*, meaning 'grey earthy rock'. By the early 1800s this limbo category 'for the reception of everything that was ancient or obscure in the geology of Britain'[1] had become one of the great challenges facing the developing science of geology.

Amid all this obscurity there appeared to be one coherent and identifiable bed of stone that offered a clue to the meaning of the Greywacke: the band of limestone running through the Gelli-grin. But as some of the greatest geologists of the nineteenth century attempted to use it to unlock its secrets, tracing the faint band of limestone as it snaked its way through the rocks of north Wales like the stripe in a tube of toothpaste, they found they couldn't agree on what it was telling them.

First into the field was the Rev. Adam Sedgwick, a brilliant but troubled Cambridge University professor dogged by ill health and bouts of depression. He had an exceptional talent for reading the rocks, but owing to some mysterious inner impediment was incapable of putting his ideas down on paper. Hard on his heels came Roderick (later Sir Roderick) Murchison, a retired soldier and ambitious socialite who treated geological fieldwork as a military campaign and conducted forced marches along Wenlock Edge and central Wales with the help of copious quantities of laudanum and the bemused support of the local gentry. For Murchison, conquering the history of the rocks was akin to conquering Africa or India: a manifestation of Britain's imperial glory.

During the 1830s these improbable companions formed one of the great scientific partnerships of the nineteenth century. Together they painstakingly mapped large areas of the Welsh Greywacke and followed the strata south as they dipped under the Severn Estuary and into Devon and Cornwall. But at the Gelli-grin they hit a band of rock that, for all its apparent clarity, they simply couldn't agree upon. Their difference of opinion – at first a hairline fracture so subtle that neither really noticed it – became increasingly acrimonious. The collaboration, initially so fruitful, collapsed, turning friend against friend, colleague against colleague, and for the next thirty years divided the Victorian scientific community.

Resolution had to wait until the 1860s, when a young provincial school teacher and amateur geologist called Charles Lapworth began his own investigation into the ancient rocks. He started tracing a little-understood family of fossils known as graptolites across the hills of the Scottish borders. His meticulous examination of the graptolites enabled him to definitively classify the Gelli-grin limestone and, in doing so, finally make sense of the Greywacke.

As I sat in the cool green shade of the Gelli-grin one summer day, with a chainsaw buzzing far away in the valley below, I became caught up in this remarkable story and its curious cast of characters. And so I decided to reconstruct their journeys around some of the most rugged parts of Britain.

I climbed the bare slopes of the Berwyn Hills with Adam Sedgwick and followed him west across the empty expanses of the Denbigh moors and into Snowdonia. I tracked Roderick Murchison along the Wye Valley and through the Welsh Marches into the green, gently rolling hills of central Wales. And I travelled north to the Scottish borders where Charles Lapworth crawled, sometimes literally on his hands and knees, in pursuit of graptolites.

As I walked in their footsteps I found myself entering the strange, almost entirely masculine world of a remarkable group of scientific pioneers. I became familiar with Sedgwick's curious blend of bonhomie, hypochondria and religious puritanism, Murchison's ruthless ambition and love of king and country, and Lapworth's patience, humility and occasional mental fragility. What amazed me was that, apart from a handful of invaluable academic works, their story has never really been told before. Yet it is a deeply involving saga of friendship and rivalry, success and failure, courage and ambition.

I also found myself entranced by the mysterious power of the rocks: so prosaic and unremarkable in their appearance, yet unimaginably old. Most nineteenth-century observers, influenced by biblical versions of the history and origins of the Earth, put their age at perhaps 5,000 years.[2] Better-informed geologists guessed at many thousand years more. But we now know that the Gelli-grin limestone is about 450 million years old, while some of the neighbouring sandstones go back another 100 million years. In the world of stones, time is measured not in years or even hundreds of years, but in hundreds of millions of years. I once picked up a very ordinary piece of slate on the slopes of Snowdon and found it utterly incomprehensible that I was holding something more than 400 *million* years old: an unremarkable object with an astonishing history.

Moreover, in the last hundred years these minerals have disclosed a record of events whose grandeur defies belief. For millions of years the land mass that is now Wales sat on a shallow coastal shelf somewhere south of what is now Cape Horn, on the borders of the Antarctic Ocean. Underwater volcanoes spewed out lava across the sea floor, sometimes shaped by the pressure and movement of the ocean currents into soft pillowy forms, at other times burying itself between layers of mud and sand. Occasionally the eruptions were so violent that they broke the surface of the water, sending clouds of dust and ash into the air, which rained back into the sea, forming yet more beds of volcanic ash among the sediments of the seabed.[3] Then things quietened down, the tectonic plates moved and 'Wales' travelled north towards the equator at the speed of a growing fingernail. The land sank under a shallow sea that alternately left beds of calcium-rich limestone such as the Gelli-grin and the muddy sandstones of the Berwyn Hills.

This is the early history of much of what now constitutes the British Isles. It's a sequence of events that Sedgwick, Murchison and Lapworth might have found hard to credit. Yet their determination to classify the Earth's early rocks laid the foundations for much of what followed.

Today, thanks to their efforts, we know the Greywacke can be broken down into a variety of different rocks: the fragments of volcanic ash and lava that cascade down the slopes of Snowdonia; the grey pebbly sandstones of the bleakly magnificent Rhinog Hills that run down the coast of the Irish Sea; the muddy silts and shales of the Berwyn Hills; and the magnificent red sandstone peaks of the Brecon Beacons.

And in breaking down the Greywacke, these geologists also gave us four distinct periods in the Earth's early history: the Cambrian, Ordovician, Silurian and Devonian, a stretch of around 150 million years during which entire continents formed and re-formed, new species of life appeared, thrived and vanished, and our world changed beyond recognition – shifting, shaping and evolving into the world as we find it today.

Chapter One

THE ERRATIC PROFESSOR SEDGWICK

At the end of July in the damp summer of 1831 a self-taught Cambridge University geologist called Adam Sedgwick climbed into a horse-drawn gig and set off from his rooms in Trinity College on the long journey to north Wales. For some years he had been casting curious glances at the region's poorly mapped mountains, but as he rattled through the lush summer countryside past banks of yellow flowering gorse, he worried that he had left it too late and would run into bad weather. Only the year before, he had been forced to abandon a trip to the Welsh hills in the face of unceasing rain.

After a day of jolting over the rutted roads, with the gig periodically engulfed in swirling storms of dust, Sedgwick stopped that evening in the booming Midlands town of Dudley.[1] At a limestone quarry near the town centre quarry men had stumbled on one of the richest fossil beds in the British Isles: a treasure trove of hundreds of marine species, many never found anywhere else in the country.

The sheer number of species suggested that this had once been an exceptionally busy underwater reef.[2] Among the unique fossils was an unusually large trilobite called *Calymene blumenbachii* – an extinct, marine, insect-like creature – known locally as the 'Dudley Bug', an aristocrat of the early seas so striking it was proudly displayed on the town's coat of arms right up until the local image was sharpened up in the 1970s. Relics like this had turned Dudley into a national centre of the fossil trade. Impromptu 'rock shops' advertised trilobites, crinoids or 'sea lilies', corals and shells for sale,[3] and Sedgwick spent a pleasant day buying specimens for the Cambridge University collection.[4]

The next morning he pushed on west across the Shropshire plain, until the road crested a wide ridge and, in a magic moment, the hills of Wales and the border country sprang up on the horizon, range after range as far as the eye could see: the Breidden Hills, the Berwyns, the Arenigs and, beyond them all, the wild emptiness of Snowdonia.

In Shrewsbury, Sedgwick bounced along wide streets of handsome buildings, the cobbles worn and polished by the passage of traffic, to a large Georgian house on the northern outskirts. He was in town to pay a visit to a certain Charles Darwin, back from Cambridge, twenty-two years old and waiting to join HMS *Beagle* as an enthusiastic but inexperienced geologist on an expedition to South America.

By way of preparation for this voyage, Darwin had offered to help Sedgwick on his tour of the Welsh hills.[5] Sedgwick was particularly well received by Darwin's three unmarried sisters, who welcomed the 'attractive and fascinating' visitor.[6] But after a day spent exploring the road cuttings and quarries around Shrewsbury the two men pushed on for Wales. Time was short and Sedgwick, perhaps flustered by all the female attention, was in a hurry to get to the hills.

In the early 1800s geology was one of several newly emerging fields of enquiry dedicated to exploring, labelling and cataloguing the natural world. The Zoological Society of London had recently opened a zoo in London to house the greatest collection of animals in the world. The Royal Horticultural Society was about to lay out Kew Gardens on the western fringe of the city, to study and display one of the world's biggest plant collections. And in the wake of the Industrial Revolution and its frenzied hunt for minerals such as coal, iron and limestone, the recently established Geological Society of London had begun identifying and mapping Britain's rocks, grouping them into simple, easily understandable categories.

The Earth had begun, they assumed, in a 'fiery baptism', which had laid down a series of hard, weather-resistant rocks. The heat and fury of their birth had turned their minerals into crystals that were usually clearly visible to the naked eye. These granites, basalts and gneiss could be found across the western extremities of Britain, where they appeared to form the core of ancient mountain ranges. They became known, for obvious reasons, as Primary rocks, the foundation on which everything else rested.

Vast ages later ancient seas had covered these rocks in water, leaving a succession of sediments that formed beds of chalk, limestone, sandstone and coal. These became known as the Secondary rocks: neat bands of strata, one on top of the other, that lie across much of England.

Finally, resting on top of them all were the sands and gravels found in areas such as Norfolk, Suffolk, Essex and the Thames Estuary, which had not yet been compressed into 'true' rock. These were the so-called Tertiary strata.[7]

This simple scheme gave early geologists three distinct categories in which to place the Earth's rocks, and a crude understanding of the temporal relationship between them. There was, of course, no way of knowing their absolute ages, so they were usually represented in a 'stratigraphic column' – a tabular representation of their relative ages, with the oldest at the bottom.

BASIC CATEGORY	COMMON NAME, 1830	MODERN NAME/ EQUIVALENT
TERTIARY	Tertiary mud, sand and gravel	Tertiary and Quaternary
SECONDARY	Cretaceous Chalk	Cretaceous
	Jurassic or Oolitic Limestone	Jurassic
	New Red Sandstone	Permo-Triassic
	Carboniferous Coal-bearing strata	Carboniferous
	Old Red Sandstone	Devonian
TRANSITION ROCKS	Greywacke	Silurian Ordovician Cambrian
PRIMARY	Primary/Primitive	Precambrian

1.1 A simplified stratigraphic column of circa 1820, showing the relative position of the different rocks found across Britain.

During the early years of the nineteenth century a proliferation of guides and handbooks, many going into multiple editions, presented these new ideas to an avid lay readership. Yet there was a problem at the heart of their simple tripartite classification, and it threatened the coherence of its whole story of the Earth.

The problem concerned a great jumble of grey, gritty sand and mudstone blocks, sometimes interleaved with volcanic ash and thin veins of limestone, that lay between the Primary and Secondary strata. This peculiar stone had nothing in common with the crystalline Primary rocks, yet it appeared to contain none of the easily discernible bands of strata characteristic of the Secondary rocks. Indeed, in some places there were no visible strata at all; in others, it looked as if they had been smashed by a giant hammer, sending thousands of fracture lines running in all directions.

In continental Europe these 'obscure heaps of sediment', as one commentator called them,[8] covered large areas of Scandinavia, the Rhineland and Bohemia, where they went by the German mining term *Grauwacke*, or 'grey, earthy rock'. In Britain they were found in parts of Devon and Cornwall, Wales, the Lake District and Scotland. The German name was anglicised to 'Greywacke', sometimes with an emphasis on the last e. They were also known as Transition rocks, because they appeared to lie at that point in the stratigraphic table where Primary rock 'transitioned' into the Secondary.

To nineteenth-century geologists and mining engineers the Greywacke rocks were so 'hardened, squeezed and broken as seemingly to defy all attempts to classify them'. 'If any one had proposed to apply to this puzzling old Transition or Grauwacke series the same tests by which the secondary and tertiary deposits had been brought into such clear and intelligible order,' wrote one observer, 'he would have raised a smile among his geological friends.'[9] They were 'an unknown land, a pathless desert', a limbo category 'for the reception of every thing that was ancient or obscure in the geology of England,' wrote another, that would require 'such a labour and devotion, in order to effect any important results' as to deter most investigators.[10]

In 1815 the land surveyor William Smith published the first geological map of Britain. It showed in considerable detail how most of the country divided neatly into Secondary and Tertiary strata: the sands and gravels of East Anglia, the chalk of the Chiltern Hills, the clays and sandstones of the Midlands, the limestone hills of the Cotswolds and the coal-bearing seams across the Midlands and northern Britain. But to the west and into Wales the neat divisions vanished and the entire area was loosely depicted as a jumble of sandstone, volcanic lava, granite and slate. Smith labelled this morass 'Red and Dunstone' – an umbrella term for the blend of sandstone, volcanic lava and granite that covers much of the region – and 'Killars and Slate', 'killars' being a Cornish mining expression for crushed and folded slates.

Smith wasn't alone in fudging the matter. The influential *Outlines of the Geology of England and Wales*, published in 1822, described the Secondary and Tertiary rocks in detail, but stopped entirely when it reached the Greywacke.[11] The equally popular *Geological Manual*, published by the West Country geologist Henry De la Beche in 1831, said of it, 'when we consider the extremely perplexing character and the geology of many of the districts where these old rocks occur, we cannot wonder that they should have continued to be a stumbling block in the progress of science ... [it would be] a hopeless task to reduce the apparent chaos to order'.[12] And in the same year Charles Lyell, one of the great Victorian theoreticians of geology, published the first edition of his enormously successful *Principles of Geology*, in which he devoted 300 pages to the Tertiary muds, sands and gravels of eastern England. The Greywacke, by contrast, got twelve lines.[13]

And yet, in so far as the fossil record could be understood, it suggested that these mysterious rocks might contain the first evidence of life on Earth. To a nineteenth-century audience deeply curious about the age and origins of life, and increasingly doubtful about the literal truth of the Bible, that gave the Greywacke an added significance.[14] If it could be mapped and logged in the same detail as the Secondary and Tertiary rocks, then geologists might begin to unravel the miracle of life itself. The Greywacke was a challenge waiting for a champion.

Sedgwick had been contemplating it for some time. He'd

grown up in the small village of Dent in the Yorkshire Pennines, rambling, hunting and fishing across a landscape of wild, empty hills and fast-flowing rivers that were remarkably similar to the Greywacke uplands of parts of Britain. It was a type of countryside he knew well and was almost instinctively curious about. Moreover, several years earlier he had struck up a correspondence with an Oxford-educated gentleman geologist called William Conybeare, a leading light in the Geological Society and joint author of the best-selling *Outlines of the Geology of England and Wales*. Conybeare had always planned a second volume on the Primitive and Transition rocks, but soon after the publication of the first volume, his co-author William Phillips, a London printer and bookseller, died unexpectedly, leaving Conybeare in need of a new collaborator. Sedgwick had agreed to step in.

A start had already been made. Sedgwick had traced the Greywacke rocks in the Lake District and had superficially surveyed the rocks of Devon and Cornwall.[15] But nobody had even attempted to analyse their composition or strata in any detail, and the Welsh hills, one of the most important areas of Transition rock in Britain, remained virtually unexplored. This was the task that was now taking Sedgwick to north Wales: to unravel the tangled knot of obscurity at geology's heart and carve out an area of expertise that he could call his own.

The journey from Shrewsbury to the Welsh border is roughly 40 miles. By midday on 5 August 1831 Sedgwick and Darwin had reached the Welsh slate-mining village of Llangollen in the Dee Valley. A river cascades through the town, passing below an old stone bridge in a series of seething rapids and whirlpools. High above the rooftops of the houses are the ruins of a thirteenth-century border castle called Castell Dinas Bran. And behind this stand the limestone cliffs of the Eglwyseg escarpment – pronounced 'Egg-lui-seg' – rearing hundreds of feet in the air and running north for mile after mile in an almost unbroken chain of crags, like a prehistoric version of the Great Wall of China.

Sedgwick had brought with him the latest map of the geology of Britain, completed only ten years earlier by George Greenough, a former president of the Geological Society. The map had been completed with the assistance of local geologists across the country and was considered the most accurate and complete chart of the rocks of Britain then available. It suggested the Eglwyseg escarpment was composed of limestone from the Carboniferous period sitting on top of a bed of Old Red Sandstone, which in turn rested directly on the Greywacke. As their gig clattered into town, Sedgwick and Darwin agreed that this was the ideal place to start mapping the Transition strata. They would log the point where the Secondary rocks gave way to the Greywacke and then travel west through Wales, moving back in geological time and charting the changing strata until they reached the even older Primary rocks along the coast of the Irish Sea. Indeed, a single journey from Llangollen to Anglesey should enable them to map the entire history of the Greywacke.

That afternoon the two men toiled up a steep zigzag footpath of loose stone and shale to Castell Dinas Bran, then stumbled across slopes of slate and scree to the foot of the Eglwyseg escarpment. Sedgwick was carrying an assortment of geologist's hammers, some with long wooden handles and heavy metal heads, and every now and then he would stop to knock off the weathered outer layer of a rock and inspect the freshly exposed surface through a hand-lens: this was really the only way to be sure of the identity of many types of rock. He may also have dripped dilute hydrochloric acid onto the broken surface: it fizzes violently in the presence of calcium carbonate and is an instant identifier of limestone.[16]

As they climbed, the two men collected specimens of typical Carboniferous fossils, including mussel-like brachiopods (also known as 'lamp shells' because of their resemblance to oil lamps) and several varieties of tube-shaped branching coral, which they carefully placed in leather specimen bags. Darwin struggled with the geology, but it's said that Sedgwick's encouragement made him 'exceedingly proud' and they worked well as a team.[17] After several hours of searching, however, it was clear to Sedgwick that they had hit a problem: no matter where they looked there was absolutely no sign of the Old Red Sandstone. The Carboniferous Limestone strata sat directly on top of the Greywacke. As one Victorian writer explained, the geological record was 'thus rendered as imperfect as an historical narrative would be if several important chapters were torn out of it and destroyed'.[18] How could Sedgwick map the Greywacke if he didn't have a clear starting point? How many other unread chapters at the top boundary of the Greywacke might also be missing?

The following day the two men climbed out of the Dee Valley on the Llangollen-to-Ruthin turnpike, tracing the line of the Eglwyseg escarpment north towards the coast. Again Greenough's map clearly showed outcrops of Old Red Sandstone lying directly below the limestone. Again it was misleading. They followed the valley of the River Clwyd, their gig trailing dust and mud through settlements of small whitewashed cottages busy with the noise of dogs, geese and bare-footed children. Occasionally they separated, taking different routes in the hope of running across outcrops of the elusive 'Old Red'. To their east the Carboniferous Limestone crags of the Eglwyseg escarpment marched across the landscape, before reaching the coast and plunging hundreds of feet into the sea in a series of spectacular cliffs at the Great Orme, an ancient centre of copper mining. 'Very obscure stratification,' noted Sedgwick as he rested in the impoverished fourteenth-century walled town of Conwy. He would later observe, 'The Old Red all round by Orm Head etc etc is a pure fiction, at least I can't see a trace of it. There is not a particle of it between Denbigh and the Isle of Anglesey.'[19]

A vital piece of the geological record – what we now know to be tens of millions of years of the Earth's history – appeared to be missing. The rocks of north Wales floated in what has been memorably described as a 'temporal vacuum'.[20] 'No amount of labour or skill,' wrote the Victorian geologist Archibald Geikie some years later, 'could possibly connect the history of the Transition rocks with that of the younger strata by which they are covered, for a great gap occurs there in the geological record.'[21]

Frustrated by this puzzling circumstance, Sedgwick decided on a new approach: he would move west. If he could not work down from the Secondary rocks into the Greywacke, perhaps he could work up from the Primary beds in Anglesey and along the north-west coast of Wales. Once again, the task looked simple enough on paper. The multicoloured, crystalline Primary rocks, their crystals catching the sunlight, would stand out against the dull brown, gritty sands and silts of the Greywacke, making it relatively easy to identify the boundary between the Primary and Greywacke strata. Sedgwick's search would switch from the limestone hills and escarpments of Llangollen, the Vale of Clwyd and the Great Orme to what he described as the 'ancient slates and porphyrites' of Snowdonia,[22] 'porphyry' being the Victorian term for crystalline, or igneous, rock.

The geologists followed a steep, stony track south along the Conwy Valley and then climbed up through the treeline and onto the bracken-covered slopes of the surrounding hills. It was hard going; Sedgwick complained of his limbs being 'jostled out of their sockets'.[23] When the going got too difficult, they abandoned the gig and continued on foot or horseback, stumbling between outcrops of rock and skirting patches where the path disappeared into brown, peaty bogs and the cold gleam of water reflected the sky.[24] On the slopes above Bangor, with the Menai Straits floating in the distance, they found a jumble of dark sandstone mixed with crystalline patches of feldspar and quartz. But still there was no obvious boundary between the rock types.

From Bangor they moved south to the grey slate quarries of Bethesda, with their chaotic hills of collapsing spoil heaps left behind by generations of miners. Several days later they took Telford's recently opened suspension bridge, vaulting 100 feet over the treacherous waters of the Menai Straits – 'the most spectacular feat of engineering ever accomplished in Britain' – to Anglesey.[25] At inlets along the coast, and in freshly worked quarries and road cuttings across the flat interior of the island, they found outcrops of unmistakable gneiss and schist. Here, intense heat and pressure had lined up the minerals into extraordinary coloured bands: the pale-grey crystals of quartz and feldspar standing out against the darker bands of mica. These were everybody's idea of quintessential Primary rock, and meant that somewhere between Bangor and Anglesey they had crossed out of the Greywacke.

Back on the mainland, confounded about where to draw the boundary, the two men separated. Darwin returned to Shrewsbury, where he would find the famous letter confirming his place on the *Beagle* and the scientific expedition that would change his life. He would later acknowledge his Welsh excursion as the origin of his interest in 'the noble science of geology'.[26]

Sedgwick, meanwhile, took off on a less glamorous four-week tour of north Wales, 'cracking the rocks of Caernarvonshire'.[27]

In the small world of early-nineteenth-century British geology, Adam Sedgwick was already a significant figure: indeed, he was one of only three paid academic geologists in the country.[28] In Italy, Germany and France the new science of the rocks was a firmly established academic discipline, but while it had spawned an army of British mining engineers and land surveyors, it was taught at just three British universities. Even there it was regarded as an 'optional extra' for students looking to widen their understanding of the natural world: 'cultural top dressing for young men preparing themselves for the traditional learned professions', as one observer witheringly put it.[29]

Geology's marginal position was reflected in its status. It lacked tradition or prestige, which meant that it was a career open to bright young men from modest backgrounds. Sedgwick, the son of an Anglican vicar in the Yorkshire Pennines, had won a scholarship to Cambridge to study mathematics, and had diligently climbed the career ladder to become a teaching Fellow, a post that saw him 'grinding away' his time with uninterested undergraduates. Sedgwick hated it.[30] The terms of the Fellowship also obliged him to become an ordained priest in the Church of England, an institution deeply enmeshed in university life at the time. As a result, Sedgwick had become a curate in the tiny village of Shudy Camps in the flatlands south of Cambridge, where he officiated at weekend services. This left him little freedom to develop his own geological speculations. 'Under such circumstances,' he grumbled, 'it is impossible to advance one step.'[31]

In 1818, however, a piece of good fortune fell into his lap: the post of Woodwardian professor of geology became vacant. The position was not so grand as the title suggests; a hundred years earlier an eccentric Cambridge naturalist called John Woodward had bequeathed the university one of the most important collections of fossils in Europe, plus land worth £150 a year, to establish a professorship in the scarcely recognised subject of fossils – what we would now call palaeontology. The gift had left the Cambridge authorities scratching their heads: should they regard it as a serious new area of scientific investigation or a niche topic of general knowledge? They concluded that it was the latter, and entrusted the appointment of the professorship to a panel including, strange as it sounds, the archbishop of Canterbury, the bishop of Ely and a local Member of Parliament. These worthies, in the absence of clear academic criteria, had decided to appoint on the grounds of what they called 'general merit'. A hundred years later nothing had changed: merit still took precedence over expertise. 'I had but one rival,' Sedgwick would later recall of his appointment, 'and he had not the slightest chance against me, for I knew absolutely nothing of geology, whereas he knew a good deal.'[32]

The salary was £100 a year (roughly £6,000 today) and meant that the holder needed private means or an additional source of income to survive. Sedgwick fell back on his post at Shudy Camps, which carried a salary of £50. But the Woodwardian professorship's conditions of employment were additionally punitive: the incumbent was expected to remain unmarried, 'lest the care of a wife and children should take the lecturer too much from study'.[33] This stipulation had led at least one previous incumbent to resign, and others had kept their personal lives well hidden.[34] Never the less, as one of only a handful of paid posts in a subject of growing importance, the role was also an opportunity. 'If I succeed I shall have a motive for active exertion in a way which will promote my intellectual improvement and I hope make a happy and useful member of society,' Sedgwick wrote to a friend.[35]

The duties were anyway hardly demanding. There was no commitment to teaching, research or departmental duties;[36] the holder was merely expected to supervise and expand the university's collection of some 10,000 fossils and deliver four lectures a year on the general topic of rocks and fossils 'to such curious persons or strangers as are engaged in the study of them'.[37] In Cambridge it was widely regarded as a sinecure and Sedgwick plunged into it. After years of what he regarded as 'dull uniformity', he had a purpose at last.

He joined the Geological Society, struck up a vigorous correspondence with amateur 'gentlemen geologists' across Britain and tentatively began to explore the country's rocks. He was a natural and charismatic speaker and his lectures, entirely voluntary and of no practical value to students in their exams, were well attended. 'Brim-full of humour and bristling with apposite anecdote, his scientific talk was greatly more entertaining than the ordinary conversation of most good talkers,' recalled one former student. 'He could so place a dry scientific fact as to photograph it on the memory while at the same time he linked it with something droll or fanciful or tender, so that it seemed ever after to wear a kind of human significance.'[38] Sedgwick would later tell his students, 'I cannot promise to teach you all geology. I can only fire your imagination.'[39]

By the late 1820s Sedgwick had become a respected figure among the small circle of men who made up British geology, and in 1829 he was elected president of the Geological Society, arguably the most prestigious post in the field. A picture of him from the period shows a determined and energetic young man with a wide brow, prominent nose and a tangle of dark hair. His build was 'spare, athletic and capable of enduring much fatigue,'

1.2 Adam Sedgwick at Cambridge in his thirties, circa 1820s.

wrote a contemporary. 'His complexion was dark – a peculiarity inherited from his mother– and his face was always wrinkled with deep lines in it.'[40] In the genteel world of Cambridge, Sedgwick enjoyed playing the bluff, no-nonsense Yorkshireman. You 'would have laughed had you seen me enter the Senate House this morning at eight,' he wrote to a friend. 'I had a cap, bands, gown and cassock – so far all was regular – but under my silk petticoats an enormous pair of mud boots and I had a great woollen ruff about my neck.'[41] Standing in front of a mirror with two friends, he once announced, 'I declare the three ugliest men in England are standing on this rug.'[42] He would later remark, 'nature cast me in a large mould, and gave me a very rugged countenance'.[43]

He revelled in the 'permission' that geology gave him to retreat for a period from the mannered world of nineteenth-century Cambridge and tramp the empty hills. 'It took him to all manner of out-of-the-way nooks and corners ... to the course of a mountain torrent, the bottom of a tree-shaded ravine, the gully cut by the frosts and rains of centuries from the face of a lonely hillside,' wrote one colleague. In the course of his work the geologist was obliged 'to double back to and fro over tracts never trodden perhaps by the ordinary tourist, but is many a time amply recompensed by the unexpected insight which this circuitous journeying gives him.'[44] Sedgwick would later observe that geology had been a 'hard task-mistress but paid me in health and happiness'. A colleague would recall that the moment Sedgwick touched 'mother earth with his hammer – down went all his fear and ill-bodings to the nether darkness'.[45]

That summer, having left Darwin with a grasp of fieldwork that would remain with him for the rest of his life,[46] Sedgwick continued on his way to Snowdon, tracing the precipitous curving path around the rim of the Glyders, a journey that adventurous tourists attempted 'with feelings of considerable alarm for it was always made out to be extremely hazardous'.[47] At Llyn Ogwen he paused to sketch a cross-section through the rocks of Cwm Idwal, where the strata form a highly visible bowl, or syncline, below Snowdon. He then travelled south towards the sandy beaches of the Llyn Peninsula and encountered more outcrops of gneiss and schist. And he climbed the steep slopes of Moel Hebog above the small village of Beddgelert, where tourists today sit in pubs overlooking the river. From here he continued south to the bleak uplands of the Rhinog grits, with their rusty-coloured sandstone terraces.[48]

Sedgwick had by now refined a procedure for mapping and logging the confusing structure of the rocks, which he had developed among the irregular and equally baffling strata of his native Pennines. He would search out a vantage point with an overview of the surrounding countryside. Often he would pause to draw the panorama; then he would look for an area of exposed rock where he could make out the lines of the strata: it might be a cliff, a small quarry or a river bank. This would enable him to identify the rock type and the angle and orientation of the strata: whether it was tilted or horizontal, and whether it ran north, south, east or west. Finally he would attempt to link one outcrop with another, so that he could determine the larger picture. If, for example, strata at Outcrop A ran roughly horizontally on a north–south orientation, and tens of miles away similar rocks at Outcrop B did roughly the same, there was a good chance the two were linked horizontally. Equally, if they rose steadily at Point A and dipped steadily at Point B, there was a strong probability there was a peak or anticline in the strata somewhere between them. In this way he could draw a series of 'traverse sections', or pictorial representations of the structure of the rocks, across miles of hillside.

But as the summer of 1831 wore on, it became clear that the rocks of north Wales were more puzzling than Sedgwick had expected. He was often confronted by an apparently chaotic mass of slates, porphyry and Greywacke, where the lines of strata appeared and disappeared amid faults and crazed, frost-driven

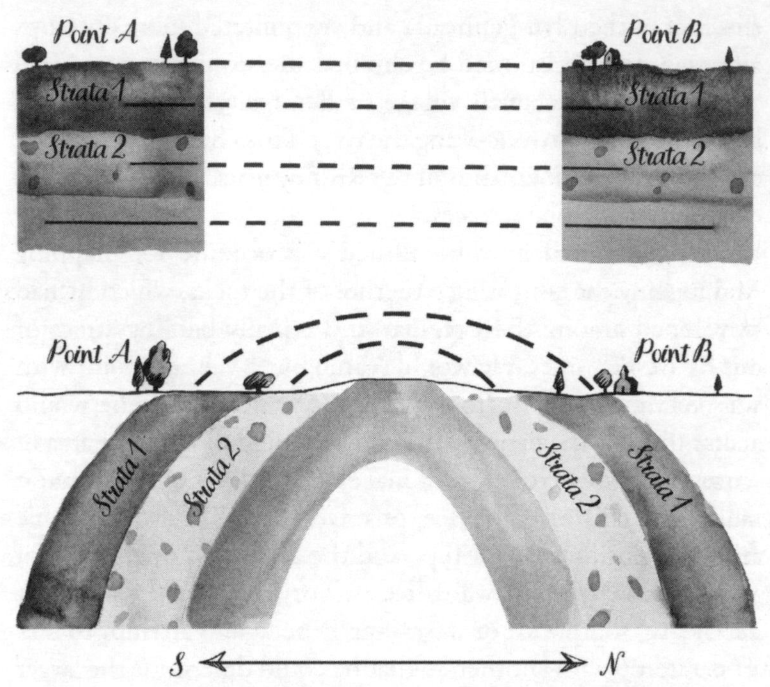

1.3 A traverse section, identifying the underlying structure of the rocks by the dip and orientation of the strata at different points.

fractures, sometimes plunging vertically into the ground, at other times disappearing altogether in beds of scree or slate. Often, as he moved from outcrop to outcrop, he had no way of knowing whether the stratum he was following was the same as the one he had seen an hour before on an opposite hill. This meant that the boundary between the Greywacke and Primary rocks, supposedly marked by a change from crystalline rocks to sandstones and silts, was far less obvious than he imagined. Hillsides of slate and scree obscured and fudged the boundary. Outcrops of volcanic ash and magma punctuated the Greywacke in unexpected places, adding to the confusion. Mapping the rocks, Sedgwick told a friend, was like 'rubbing yourself against a grinding stone'.[49]

By September he had reached Cadair Idris in the southern hills of Snowdonia, where local guides charged five shillings for a tour of the mountain and delighted in describing the hazards of going it alone. One common story told of a Mr Smith who had vanished on a September day, only to be found the following May at the bottom of a precipice, his remains chewed to pieces by foxes and ravens. 'His eyes were gone. His teeth were dashed out by the fall and lay scattered about the mountain. His head was bent double under him and crushed into his chest,' went the account. 'The only piece of flesh remaining on the bones was where the coat buttoned over the chest.'[50] For his part, Sedgwick made the climb without misadventure, but the endless days of travel were taking their toll.

He was obliged to follow 'detestable roads' that turned into muddy pools of water and bogland after periods of heavy rain. 'Everything is wrapped in mist and the rain is falling in buckets,' Sedgwick reported on one occasion,[51] and he was forced to spend long, idle days sheltering 'in the outhouse of a cottage, in a most dismal situation'.[52] He began to complain of rheumatic pains and nights racked by fevers brought on by the rigours of the day. 'Lord pity the arse that's clagged to a head that will hunt stones,' pleaded a fellow geologist.[53] Reports began to reach Cambridge that Sedgwick was 'full to the teeth with Welsh porphyry and grauwacke' and had begun to loathe the 'old spirits of the mountains who made a great resistance of the process of being geologised'.[54]

Finally, at the end of September 1831, as the university autumn term was about to begin, he gave up. He had covered a huge amount of ground and had made extensive notes. He believed he could loosely identify a baseline where the ancient

Primary rocks of Anglesey and the Llyn Peninsula gave way to the younger Greywacke rocks of Snowdonia.[55] He had also drawn dozens of idealised cross-sections through the rocks of the region, which suggested that, on the basis of the general run of the visible strata, the region was composed of a series of north–south running folds or corrugations. The sands and siltstones of the Rhinog Hills appeared to have been pushed up to form what Sedgwick would call the 'Merionethshire anticline' or the 'Harlech Dome', while the rocks of Snowdonia had been folded down at places like Cwm Idwal into a trough or syncline, which would later become known as the Central Wales Syncline. These folds, Sedgwick speculated, might repeat themselves to the west and east.

It was an impressive piece of work. I once gazed at a cliff in the Snowdon foothills and wondered how he had ever managed it. The face was covered in bracken, grass, scree and chaotic outcrops of slatey rock splintered and scarred by the weather and by movements in the Earth's crust. The strata were hard to find and almost impossible to follow. Yet in a matter of weeks Sedgwick had mapped, with what would turn out to be uncanny accuracy, the underlying structure of the rocks of the entire region.

And yet he had still failed to find a clear boundary between the Transition and Secondary rocks. He had made little or no progress in making sense of the Greywacke or in dividing it into discrete strata. Back in Cambridge, he told a colleague, 'the weather became so bad that I was driven out of Caernarvonshire before I had quite finished my work; but God willing, I hope to be in north Wales next year before the expiration of the first week in May and with five months before me, I shall perhaps be able to see my way through the greater part of the Principality.' He added, 'What a horrible fraction of a geological life

[is] sacrificed to the most toilsome and irksome investigation belonging to our science!'[56]

Further south, however, among the grassy hills of Shropshire and mid-Wales, another geologist was enjoying considerably more success.

Chapter Two

THE AMBITIOUS MR MURCHISON

During the early summer of 1831, as Sedgwick and Darwin were still planning their trip, a young 'gentleman' geologist named Roderick Murchison was already on his way to Wales. As a man of means, he travelled in style, with a 'maid, two good grey nags and a little carriage, saddles being strapped behind for occasional equestrian use.'[1] He was accompanied by his wife Charlotte, a naturalist and artist whose habitual task was to tag rock specimens and sketch the landscape. The party set off from the Murchison's town house in London's newly fashionable West End, pausing en route to visit Murchison's former tutor, William Buckland.

Buckland was Sedgwick's opposite number at Oxford, making him the only other full-time, paid, academic geologist in Britain. He was a well-known eccentric who took great pleasure in having identified the small, sausage-like fossils known as coprolites as the remains of dinosaur droppings.[2] 'I entered a long corridor like room filled with rocks and bones in dire confusion,' Murchison wrote of a visit to Buckland's rooms in Oxford. 'In a sort of sanctum at the end was my friend in his black gown looking like a necromancer, sitting in a rickety chair covered with some fossils, clearing out a fossil bone from the matrix.'[3] Murchison talked of his plans for a general tour of the Welsh rocks. Buckland emphasised the difficulties of the country's puzzling geology, but failed to dent his former pupil's enthusiasm.

The following day the party moved on from Oxford, climbing over the grassy southern flanks of the Cotswold hills before dropping down into the wide, flat expanse of the Severn Valley. They crossed the river by ferry somewhere west of Gloucester, weaving between horse-drawn barges and the billowing sails of inshore wherries, and pushed on to the small fishing village of Sully on the banks of the Severn Estuary, today a suburb on the southern fringes of Cardiff. This was the home of local vicar and fellow gentleman geologist William Conybeare: the very same man who was collaborating with Sedgwick on the Greywacke project.

Murchison in fact knew both Conybeare and Sedgwick rather well: they all encountered one another regularly at meetings of the Geological Society in London. There had even been talk of Murchison joining Sedgwick on the Welsh expedition. But Sedgwick's departure from Cambridge had been delayed, and Murchison grew impatient. He only had a passing interest in mapping the Greywacke, in any case; his principal concern lay elsewhere.

Murchison never planned to be a geologist. He was the son of a wealthy family of minor Scottish landowners and had always dreamed of military glory. He was sent to the Royal Military College at Great Marlow in Buckinghamshire at the age of thirteen,[4] joined the army at sixteen and, within a year, left for Spain and the Peninsular War, in which Britain, Spain and Portugal were challenging Napoleon Bonaparte's ambitions to expand the French Empire into south-west Europe. The campaign, however, went badly: within six months of his arrival, the British were forced out of the region and Murchison was back in Britain.

It was not the life of military glory he had dreamed of, and he fervently hoped the continuing war with Napoleon would offer fresh opportunities to return to the front.

It was not to be. In June 1815 the Duke of Wellington defeated the French at Waterloo and peace returned to Europe. To most of the country this came as a welcome end to a war that had dragged on for twelve years. Half a million young men were called up to fight, thousands never returned; not for nothing was the war known as the 'meat grinder'.[5] A vivid account of news of the victory comes, strangely enough, from Adam Sedgwick, who had left Cambridge that summer to escape a typhoid outbreak and was cooling his heels at his parents' home in Dent. To pass the time he took to riding over to the neighbouring town of Sedbergh to collect the village post and newspapers. About 23 June his arrival coincided with the delivery of an extraordinary edition of the *London Gazette*, the government's official journal of record, carrying news of the victory at Waterloo. 'After joining in the cheers and congratulations', Sedgwick noted:

> I returned to Dent with what speed I could; and such was the anxiety of the day that many scores of my brother Dalesman met me on the way, and no time was lost in our return to the market place of Dent ... Then, mounting on the great blocks of black marble, from the top of which my countrymen have so often heard the voice of the auctioneer and the town crier, I read, at the highest pitch of my voice, the news from the Gazette Extraordinary to the anxious crowd which pressed around me. After the tumultuous cheers had subsided, I said: 'Let us thank God for this great victory, and let the bells give us a merry peal.'[6]

It was a blow for Murchison, though. He was still in his early twenties, a captain on half-pay, and peace in Europe dramatically reduced the possibilities of military advancement. The Battle of Waterloo, he later told friends, 'submerged all my ambition as well as that of the great Napoleon'.[7] He resigned his commission and two months later married Charlotte Hugonin, the daughter of a wealthy army officer. The couple moved to Teesdale in the hills of northern England[8] and then to rural Leicestershire. Murchison gave himself up to a playboy life of entertaining, partridge-shooting and fox-hunting. For months on end, runs one account, bored with the 'insufferable ennui' of paying social calls, he rode with the hounds almost daily, becoming one of the most celebrated fox-hunters in the north of England.[9]

Murchison managed five years of this extravagant existence before it became clear that he was living beyond his means. Subscribing to a pack of hounds, running a stable of horses and entertaining on a lavish scale all turned out to be unexpectedly expensive.[10] Looking to cut costs, the Murchisons sold up and moved to a smaller (if still handsome house) in London. But what was he supposed to do with himself there?

Several years earlier, at a partridge shoot in the Leicestershire wolds, Murchison had run into the eminent chemist Sir Humphry Davy, who spoke with infectious enthusiasm of the new and expanding world of scientific investigation. 'As we shot partridges together in the morning,' Murchison later recalled, 'I perceived that a man might pursue philosophy [that is, science] without abandoning field sports.'[11] He began to attend lectures at the Royal Institution, an organisation devoted to scientific education and research,[12] encouraged by his wife Charlotte, who is said to have quietly worried that he was dissipating his 'not inconsiderable abilities in hedonistic frivolities'.[13] And so,

2.1 Roderick Murchison in 1836 when he was in his early forties.

for want of anything better to do, Murchison joined the Geological Society and studied in Oxford under William Buckland.

He might have felt rather at home. The study of the Earth's rocks may not have been viewed in Britain as a serious field of academic study, but for this reason it was widely regarded as 'a vocation worthy of a gentleman offering intellectual satisfaction and social status commensurate with the established professions'.[14] It meant the upper ranks of the geological establishment were stuffed with 'affluent, ex-military, rural squires',[15] who sought in geology fieldwork at least some of the adventure of a military life and the outdoor pleasures of fox-hunting and country sports. Such men enjoyed some technical advantages, too: the practice of 'military drawing' – the process of quickly surveying and sketching the lie of the land – was considered an essential skill without which, in the words of one practitioner, 'the geologist will often find himself much embarrassed'.[16] Murchison would tell a friend he hoped to demonstrate 'that a gentleman of science could wield both a hammer and a gun in the field' and that 'a sporting man could also be a scientific one'.[17]

Not everybody was convinced he would pull it off. 'He is a diligent observer & I believe a good naturalist but does not appear possessed of much originality,' observed a fellow Scot.[18] Yet for many people originality and talent were not regarded as high on the list of requirements for a study of rocks; it was generally assumed that good work could be done by a man with a quick eye, a stout pair of legs, good judgement and a clear notion of what had already been accomplished. What Murchison may have lacked in originality, he made up for in energy and ambition. Furthermore, unlike many gentleman geologists at the time, he still had funds enough to pursue his new interest full-time.[19]

The Murchisons spent several days with the Rev. William Conybeare and his family at the rectory in Sully. There is no record of their conversation, and Murchison's field notes for the period, lodged in the tall, graceful library on the first floor of the Geological Society's premises in London's Piccadilly, are confined to brief accounts of visits to local sites of interest and offer little insight. Nevertheless, it would have been surprising if they didn't talk about the enigmatic Greywacke, one of the great challenges of the age, and the significance of Conybeare's collaboration with Sedgwick in trying to disentangle it. But Murchison almost certainly had his own, rather different reasons for wanting to consult Conybeare on the Welsh rocks.

The first was a cartographical issue. George Greenough's geological map of Britain, as Sedgwick and Darwin were about to discover further north, was known to contain important inaccuracies along the English–Welsh border, where it showed a clear, if irregular band of Old Red Sandstone running up through the region. Murchison had therefore agreed with Greenough that during his tour of the Welsh hills he would check for more accurate and detailed local information before a second edition of the map was published.

But there was also probably a second motive. For several years Murchison had been collaborating with the geologist Charles Lyell, a passionate advocate of the theories of a celebrated eighteenth-century Scottish polymath named James Hutton. Forty-five years earlier Hutton had published a radically new version of the history of the Earth, which challenged the popular biblical account. It drew on recent experiences of major volcanic events in southern Europe and in particular Iceland, where, during the summer of 1783, a massive eruption had blanketed much of Europe in a thick cloud of dust and fog for months on end.[20]

Hutton began to speculate about a restless febrile Earth with a blazing core that periodically 'boiled over', sometimes breaking through the Earth's crust in the form of volcanoes, but at other times simply heaving up the land, tilting and bending the strata and forming new hills and mountains: a process now referred to as 'crustal uplift'. Natural forces like the wind and rain would then set to work, eroding these new uplands and carrying loose rocks and soil to the sea, where they settled on the bottom, gradually creating new layers of rock, which were in turn heaved up by the boiling core: an endless cycle of uplift, erosion and deposition, which some romantics had begun to refer to as the 'symphony of the earth'.[21]

Hutton's model was one of the first attempts to outline a comprehensive theory of the history of the Earth that replaced divine intervention and supernatural forces with observable physical phenomena.[22] It became known as Uniformitarianism – the idea that the Earth is entirely shaped by slow-moving forces acting over long periods of time.[23] Murchison, new to geology and still finding his feet, had been deeply intrigued. He agreed to join Lyell in a grand project to substantiate this still-controversial and challenging theory.

For ten months in 1828 the two men toured the volcanic regions of France and Italy. In their hills and valleys they found signs of a landscape repeatedly shaped and altered by volcanic eruptions and subsequent erosion by rivers: suggestive evidence for Hutton's thesis.[24] Back in Britain, Murchison had agreed to continue his collaboration with Lyell. It was almost certainly this radical hypothesis that took him to Wales, where a small range of ancient volcanic hills in the centre of the country promised another opportunity to substantiate their ideas.[25]

As Conybeare and Murchison toured the sea cliffs west of Sully that summer and traced the coast to Swansea and its iron foundries, there is no evidence that Conybeare had anything useful to contribute to Hutton's model of crustal uplift – not that he would necessarily have been opposed to it. By the late 1820s even deeply religious clerical geologists like Conybeare and Sedgwick had come to accept that biblical accounts of the Earth's history were inadequate, and were willing to consider alternative, less apocalyptic models. All we can say for sure, however, is that Conybeare encouraged Murchison to investigate sites where the 'Old Red' reportedly sat on top of the Greywacke, and to contribute to the general improvement of Greenough's map.

Two days later the Murchisons moved on, climbing out of the coastal plain through bands of Carboniferous coal, limestone and grit as they travelled north towards the older hills of central Wales.

Their destination was a series of volcanic hills near the Victorian spa town of Builth Wells, in what was then southern Radnorshire, where there had been reports of marine fossils pushed up from the seabed, possibly by Hutton's 'crustal uplift', so that they were now lying high on the slopes of the hills.

As they journeyed north through a countryside of steep woods and isolated hamlets they encountered outcrops of Old Red Sandstone sitting in an unbroken sequence on top of Greywacke rock: the sort of clear boundary between the two rock 'systems' that Sedgwick was beginning to find so elusive further north. Murchison duly logged them; this was the kind of detail George Greenough would need for his revised map. Two days later, after a series of steep climbs and descents through rolling hills of red earth and fresh, green grass, through villages where half-clad children came out to watch their passage, the party dropped down into the valley of the River Wye. The river here runs fast and shallow, 'winding down like a silver serpent' between steeply rising wooded banks.[26]

The Murchisons followed the west bank north until they reached a point where the lane climbed past a large red-brick house known as Trericket Mill. Here, peering down across the river, Murchison noted that 'low terrace-shaped ridges of grey rock dipping slightly to the south-east appeared on the opposite bank of the Wye, and seemed to rise out quite conformably from beneath the Old Red of Herefordshire'.[27] It was another neat and undisturbed boundary where the jumble of grey mud and sandstone that made up the Greywacke passed directly up into the Old Red Sandstone. But then comes a mystifying change in tone.

'Boating across the river at Cavansham Ferry,' Murchison breathlessly recorded, 'I rushed up to these ridges, and to my inexpressible joy found them replete with Transition [Greywacke] fossils.'[28] The excitement seems out of place: after all, he had already passed several similar sequences where the Greywacke passed smoothly up into the Old Red, without any obvious sign of emotion. His old friend Buckland had even told him to expect such a sequence along the Wye Valley. Why the sudden urgency and elation?

The answer may lie in the fact that this is almost certainly not an accurate record of what happened. The account was rewritten many years later, when, as we shall see, Murchison was anxious to put a rather different spin on his first foray into the Welsh rocks. At the time it's more likely that he merely logged another undisturbed junction between the Greywacke and Old Red Sandstone, then continued north to Builth Wells, where the Wye squeezes through the graceful arches of an eighteenth-century, stone bridge. Here he climbed steeply out of the valley along narrow overgrown tracks between thickets of bright-yellow flowering gorse, towards a series of squat volcanic hills to the north-east: hill after hill appearing in the distance across open country until they vanished in a blue haze.

We can only speculate as to his true feelings at the time. Murchison was not especially susceptible to the romantic sentiments of many of his colleagues. His instincts, groomed by a childhood in the army and encouraged by military relatives, tended towards a more soldierly outlook, and he had come to regard fieldwork as a form of military campaign in which the objective was to conquer the natural world rather than admire it.[29] In language that was probably second nature, he wrote in his field notes and letters to friends of 'forced marches', 'battles', 'salvoes' and 'counter-attacks' as he 'invaded' the hills. Murchison prided himself on his physical prowess and stamina, once boasting that he had covered 120 miles in three days.[30] His approach to fieldwork 'possessed something of the character of a forced march, a "keep-moving, go-at-it-if-it-kills-you system"', his friend Charles Lyell wrote after their trip to France, 'and I had to fight sometimes for the sake of the geology, as his wife had to for her strength, to make him proceed with somewhat less precipitation'.[31]

Spartan as this sounds, Murchison also liked to have his creature comforts accommodated and, although he certainly spent many nights in damp, dreary inns, he carried carefully garnered letters of introduction to the local gentry – politicians, businessmen, local squires and contacts from his period in the army – who had a reputation for being 'exceedingly civil to strangers'.[32]

One such man would become a particularly generous and important host. Sir Thomas Frankland Lewis was the Tory MP for the local constituency of Radnorshire (now part of Powys) and the owner of Harpton Court, a large country estate straddling the English–Welsh border outside Builth Wells.[33] Sir Thomas, like many shrewd landowners, had invited Murchison to explore his land in the hope of finding newly valuable minerals, and the Murchisons spent a busy week at Harpton Court – the first of many visits – exploring the volcanic uplands of central Radnorshire near the estate with the help of a local quarry man.[34] Frankland Lewis later wrote fondly that he would never forget the confusion of maps that the geologists spread out across the floor of his study.[35] Murchison filled twenty-six pages of his notebook that week, noting that he had found no sign of useful minerals, but had unearthed a rich fossil record of marine life, including brachiopods, oyster-like bivalves and sponges: more evidence of James Hutton's cycle of 'volcanic uplift'.[36]

This was not the only significant thing Murchison stumbled upon while staying with Lewis, however. A vicar-geologist happened to be living nearby in the tiny North Herefordshire village of Aymestrey. The Rev. Thomas Lewis (no relation to Sir Thomas) had trained under Sedgwick in Cambridge, and in a later age might have become a professional geologist. But in early-nineteenth-century Britain, like many others, he had turned to the Church for a living and pursued geology on the side. In between services at the village's small Norman church, Lewis had spent four years mapping the area's rocks and zealously collecting their fossils, 'which were everywhere in abundance strewed over the roads and fields' around the village.[37] Murchison was intrigued and paid the man a visit.

The two of them took a walk through Aymestrey, with its half-timbered coaching inn on the River Lugg, and across the surrounding hills, where the earth changed from brown to red, and Lewis was able to show Murchison a small area of exposed rock where, once again, the Old Red Sandstone could be seen sitting neatly on top of the Greywacke: yet another beautifully clear example of the boundary where the two geological periods met.

But Lewis showed Murchison something much more intriguing as well. He took him home to the modest vicarage where his fossil collection was stored in a cabinet containing five drawers, each one allocated to a particular bed of rock and its associated fossils. Many came from the beds of Greywacke rock and meant that Lewis could not only divide it into a series of distinct strata but, even more remarkably, could assign a discrete fossil 'fingerprint' to each one of them.[38]

It's hard to overstate the significance of this achievement. For some years geologists had been cataloguing the fossil-rich Secondary rocks, the bands of sandstone, limestone, clay and coal that cover much of England, on the basis not only of their appearance or mineral content, which can be remarkably varied, but also their fossil record. The idea here is that each type of sedimentary rock is laid down under very particular conditions, which in turn give rise to one or more unique life forms. When these die, they leave a fossil record that is specific to that period. It might take the form of a single species, a particular abundance of a number of species or, most commonly, a unique combination of different species. In other words, all rocks formed during the same geological period ought to share the same distinctive fossil fingerprint, an indelible marker running through the apparent chaos of the strata.[39]

To Murchison it must have come as a revelation, and his field notes for the period, drawing heavily on Lewis's work, break down the Greywacke into six alternating bands of limestone and shale – a fine-grained mixture of mud and limestone – each one distinguished by its own particular characteristics and fossil content.

Murchison spent another week travelling northwards up the English–Welsh border, past fields of red-and-white cattle, following Greenough's map and attempting to trace out the boundary between the Secondary and Transition rocks. In the ancient border town of Ludlow, a centre of mineral exploration and 'one of the neatest, clean, pretty towns in England',[40] he was shown around by another local geologist: a doctor by the name of Thomas Lloyd, who took him to a site on the outskirts of the town called Ludford Corner and showed him a small cliff where, again, the junction between the Old Red and the Greywacke was plainly visible. Today the cliff is almost hidden by low tree branches, but an information board reminds visitors of Murchison's visit.[41]

Lloyd also took his visitor to a point where the River Teme flows clear and fast in a wide arc below the town and its Norman castle. On the opposite bank the Greywacke rises steeply in a limestone cliff where local geologists had discovered rich deposits of a brachiopod called *Pentamerus knightii*, a fossil that Murchison had recently seen in Aymestrey. This suggested that the Ludlow and Aymestrey Limestones probably belonged to the same stratum of Greywacke rock: what would sometimes be referred to as Pentamerus Limestone.[42] It was further evidence that at least the 'upper' Greywacke rocks were susceptible to being broken down into distinct bands on the basis of their fossil record.

As he left Ludlow and continued north to York and the inaugural meeting of the British Association for the Advancement of Science, a gathering of some of the country's leading men of science, Murchison seems to have realised that his visit to the Welsh Marches had put him in an extraordinarily strong position to address an issue at least as interesting as Hutton's notion of a restless Earth: the problem of the Greywacke.

In the grand surroundings of the Yorkshire Museum, a recently built neoclassical building in the town centre, he hung up the coloured maps he had used that summer to mark out the junction between the Secondary and Transition rocks. Pointing to these charts, he boldly announced to an audience of '300 gentlemen'[43] that they represented 'a new step in British geology', in which 'the true relations of the Transition rocks to the later formations was now at length to be dispelled'.[44]

He would later write, 'I had got the Upper Grauwacke so-called into my hands for I had seen it in several situations far from each other all along the South Welsh frontier, and in Shropshire and Herefordshire.'[45] This was an outrageous claim: all he had done in most instances was follow what others had told him. Yet to many at the meeting it must have seemed as though Murchison had single-handedly filled the puzzling 'temporal vacuum' between the Greywacke of Wales and the Secondary rocks of England. What Greenough's map had until now shown with rough and questionable accuracy, Murchison could firm up in detail.

But this was just the beginning. He went on to tell his audience that he had achieved something even more important: he had explored a section of what appeared to be upper Greywacke rocks and had not only found that it could be broken down into discrete strata, but that each bed could be 'characterized by peculiar fossils'.[46] He made no mention of the Rev. Thomas Lewis or other local experts who had guided him through the region. To most of his listeners it must have appeared as if it was all his own work.[47]

Murchison would later write of the summer of 1831 that it had been 'the most fruitful year of my life ... I was then 39 years old and few could excel me in bodily and mental activity.'[48] His sense of triumph was justified at least to some degree, notwithstanding his reluctance to credit those who had helped him. His biographer, the Scottish geologist Archibald Geikie, would later write patronisingly of men like Thomas Lewis and Thomas Lloyd that 'though not aspiring to be called geologists ... [they] had been in the habit of looking at the rocks and fossils of their neighborhood', adding that 'from the first these friends enlisted readily in [Murchison's] service'.[49] According to another account, Lewis 'cheerfully resigned the subject, rejoicing that it had fallen into the hands of a geologist whose practical knowledge was much greater than his own.'[50]

Both these statements were false. Lewis had planned to publish his findings himself and would later sorely complain that he received so little acknowledgement for his contribution to the study of the Greywacke. 'I cannot withhold from you that I felt a disappointment in the slight notice my early researches have received,' he wrote to Murchison years later:

> I cannot be ignorant of the importance of my early doings, of the accuracy of the succession I had observed of the rocks in the neighbourhood of Aymestrey, previous to your first visit to that locality, and the value of my subsequent identifications and of the richness of illustrations I there laid before you and the liberality with which I continued to supply you with everything that came within my reach.[51]

That winter Sedgwick and Murchison compared notes in London. Sedgwick had collected a mass of data about the structure and nature of the rocks of north Wales, which, through a fog of confusion, suggested that they had been heavily disturbed and folded into a series of roughly north–south-running domes and troughs. But he had been unable to link the Secondary rocks to the Greywacke or to identify any distinct strata in the older rocks. As a result, his project to map the Transition rocks of Wales was still in its infancy.

Murchison, meanwhile, appears to have quietly changed his plans. His original idea had been to continue the collaboration with Charles Lyell. He now realised that he had stumbled on a new and much more exciting project. For the first time in his life he was in a position to emerge from the shadows of other men and strike out on his own.[52] As the historian of science James Secord has written, Murchison had 'discovered' not merely clear examples of the boundary between the 'Old Red' and the Greywacke, but 'rather a potential new direction for his research'.[53]

It's almost certainly in this light that we should reread his breathless account of events near Trericket Mill on the Wye.[54] It was written some thirty-five years later, when the Greywacke had become central to Murchison's reputation as a geologist and he was anxious to demonstrate that his exploration of it in the summer of 1831 had been a deliberate plan rather than a happy accident.[55] It may also have helped that it could be spun to fit the convention of the Victorian romantic hero who sets out on a journey of exploration and experiences a blinding 'eureka moment' when all is revealed, and, moreover, that it occurred before the encounter with Thomas Lewis. Murchison's biographer would later embellish the story still further, describing it as 'a find' by an 'adventurous spirit' on 'a sublime scientific quest'.[56]

Today many geologists believe the entire incident may have been fabricated. There is considerable doubt that the 'ridges of grey rock' and the junction with the 'Old Red of Herefordshire', are visible from the alleged point on the river; you have to climb perhaps a quarter of a mile up the side of the valley to see the rocks properly, and even then the junction between the Greywacke and the Old Red is difficult to find. As the geologist Duncan Hawley, who has explored these rocks in great detail, reports, 'there is a small stream in the wood and by carefully following this upstream it is possible to find the change from grey to red beds in the stream bed. The exposure however is poor.'[57]

There is even some debate about whether the junction between the Greywacke and Old Red Sandstone is visible at all on this hillside. Nowadays the ridges are forested, but in Murchison's day the area was extensively quarried and there was plenty of bare rock. But according to some observers, the junction between the Greywacke and Old Red is not in this area of once-bare rock at all, but higher up in a field.[58]

Whatever the truth, during the winter of 1831–2, Murchison, at the age of thirty-nine and only seven years after he had taken up the subject, was handed one of the greatest accolades of British geology: he was elected to succeed Sedgwick as president of the Geological Society.

It was now that Sedgwick and Murchison decided to form what would become one of the great scientific collaborations of the age: a grand project to untangle and map the Transition rocks so that they could be brought into line with the rest of the stratigraphic table. They agreed that Sedgwick would continue to concentrate on north Wales – from the Berwyn Hills to the mountains of Snowdonia and the north Wales coast – while Murchison would explore the area he was already familiar with: southern Shropshire, the Welsh border counties of Radnor and Montgomery and south and western Wales.

The division was largely arbitrary. The Greywacke rocks of Shropshire and the Welsh borders clearly lay just below the Old Red Sandstone at the base of the Secondary strata, while the beds in north and western Wales seemed to be closer to the very ancient Primary rocks of Anglesey and the west coast. This suggested that the rocks north and west were probably older than those further south. Indeed, the rocks themselves seemed to support this assumption: the Shropshire Greywacke contained a rich fossil record that pointed to a flourishing early marine life, while the record further north was much sparser, suggesting that life on Earth had just got going. There appeared, in other words, to be an 'upper' and 'lower' Greywacke.

2.2 Sedgwick and Murchison's division of the rocks of Wales in 1832.

It was not, at the beginning, a collaboration of equals. Murchison, still unsure of his geological skills, appears to have felt the need to hitch his fortunes to someone he regarded as more experienced. In the 1820s it had been Charles Lyell; now the duty shifted to Adam Sedgwick. 'From his buoyant and cheerful nature, as well as from his soul and eloquence, Sedgwick at once won my heart,' Murchison told a friend.[59] (Lyell would later write waspishly to a common friend, 'you know that he idolises "the first of men," as Adam is usually styled'. [60])

Sedgwick's motives for making common cause were less clear. He may even have been a reluctant collaborator at first.[61] But he appears to have come to value the younger man's energy and organisational skills. Over the months the pair grew close, their frequent correspondence moving from 'Dear Sir', to 'Dear S' and 'Dear M', until the point at which Murchison might sign off 'Rodericus' and Sedgwick would reply, 'Yours to the centre of the earth'.

Chapter Three

TREKKING THROUGH WALES, 1832–3

As you travel west towards the Welsh border, the Breidden Hills rise dramatically out of the surrounding countryside: first a smudge of distant uplands glimpsed through the trees and hedgerows, then a series of prominent peaks that dominate the surrounding farmland. These are in fact A. E. Housman's 'blue remembered hills':

> In summertime on Bredon
> The bells they sound so clear;
> Round both the shires they ring them
> In steeples far and near,
> A happy noise to hear.

Here, on a late June morning in 1832, possibly to the sound of church bells, Adam Sedgwick paused on his way to a second season of fieldwork. He climbed the tallest of the five peaks, Breidden Hill itself, following a steep path through low-lying bushes of winberries heavy with black fruit, and across areas of dark crystalline rock that twinkled in the sunlight: the remains of volcanic upheavals hundreds of million years ago.

As he neared the top, Sedgwick would have seen the slim 50-foot column of Rodney's Pillar silhouetted against the sky: the monument had been erected fifty years earlier by the

'Gentlemen of Montgomeryshire' to commemorate the victory of a local hero, Admiral George Brydges Rodney, in an obscure sea battle during the American War of Independence.[1] But Sedgwick would not have been thinking of Rodney. Standing at the foot of the pillar, he had a panoramic view across the Welsh hills from Snowdon in the north to the ghostly curved outline of Cadair Idris to the south-west. Somewhere in this wild sweep of countryside must lie the key to the secrets of the Greywacke. Then he turned south to the neighbouring peak of Moel y Golfa, where, as it happened, he could just make out a small group of figures toiling their way to the top. This was Roderick Murchison and his wife Charlotte, also conducting another summer of fieldwork in Wales, pausing to view the surrounding countryside and the task ahead.

Sedgwick continued westwards in the evening light, heading for the ancient mining village of Llanymynech, a dusty one-street settlement straddling the English–Welsh border. The local population had once scraped a living mining copper and lead in the surrounding hills, but by the 1800s the mines were almost exhausted and most men now worked in the nearby quarries and kilns. Dust and the fumes of burning limestone hung in the air. 'I landed that evening at Llanymynech, and spent next day with my friend Evans in a proper clerical manner,' Sedgwick wrote to Murchison in the first of what would become a regular series of reports that summer. 'I did, however, after church, go up on the hill north of the town, and enjoyed what I think one of the very finest views I ever beheld.'[2] This was Llanymynech Hill, which looks south over Radnor Forest, and west to the Greywacke hills of Snowdonia.

Sedgwick pushed on into Wales, past the flourishing country estates of local landowners,[3] and by mid-morning the next day his gig had reached the foot of the Berwyn Hills, rising steeply out of the Shropshire flatlands, at a point known as Pistyll Rhaeadr. Here a small stream plunges down the eastern flanks: 'The water tumbles over the Greywacke and porphyry through a perpendicular height of about 230 feet,' noted Sedgwick, adding that the huge height of exposed rock made it 'a gorgeous fall for a geologist'.[4]

He was looking once again for exposed outcrops of rock where the line of strata might provide an insight into the underlying structure. But the rocks at Pistyll Rhaeadr were composed of irregular blocks of grey Greywacke sandstone and it was impossible to make out the pattern in any detail. Up, then, through wooded slopes to the treeless peak of Cadair Berwyn, rising nearly 3,000 feet in the air.[5] Here Sedgwick looked east over the Cheshire plain, with the shadow of the Pennines trembling in the distance, but again there were no obvious clues to the histories written beneath his feet.

The following morning he climbed further into the hills, following a drover's road past slate quarries, spoil heaps and the remains of derelict lead workings. Slopes of gorse and bracken rose on either side, cut through by busy mountain torrents and precipitous falls of scree. As he climbed higher there were almost no visible rock outcrops. But near the summit, at a place known as the Milltir Cerrig pass, where a small footpath crosses the track on its way north to the Welsh coast, the land flattens out and a jumble of large displaced boulders mark the site of a small quarry cut into the Greywacke.

Sedgwick paused to look closely. The stones had been blackened by the weather, but the rough sides of the quarry suggested that the general line of the strata ran in a roughly north–south direction, following the contour of the hills. Working a rock loose and knocking off the surface with a hammer, he could also distinguish a paler line of rock threading its way through the beds of sandstone and slate: it reacted furiously with dilute hydrochloric acid, a sure sign of limestone.

He followed a steeply descending track down the western slopes of the Berwyns, with ledges of bare stone breaking through the surrounding grassland as if the bones of the Earth were bursting through its skin, towards the shores of Lake Bala. Here he paused at a second small quarry near a farm known as Gelli-grin where, among green shoots of bracken thrusting up through banks of dead leaves, he found a second irregular band of 'black shelly limestone' that bore a striking resemblance to the rocks he'd just explored at the Milltir Cerrig pass. Geologists have two useful expressions for describing bands of strata like these: the compass orientation of the strata – whether it runs north, east, south or west – is known as its 'strike', while its angle relative to the rocks below it is its 'dip'. The limestone strata at Milltir Cerrig had a roughly north–south-running strike and a north-west-facing dip. The rocks at Gelli-grin shared the same strike, but had a south-east-facing dip.

At last something to go on. Over the next few days Sedgwick traced further outcrops of these gritty shale and limestone rocks as they ran north–south down the line of the Berwyn Hills. He found that they appeared to form two roughly parallel bands, one snaking down the western flank of the hills and passing through the quarry at Gelli-grin, the other running further east and passing through the Milltir Cerrig pass. He named them the Bala Limestone, after the nearby lake.[6]

It was the first time Sedgwick had been able to match Murchison's achievement of the year before and trace out a single band of strata in what he was now calling the 'crusty old' Greywacke. What was more, he could draw several important conclusions. First, he could make a crude assessment of the relative age of different areas of rock. Everything immediately below the limestone strata – assuming it hadn't been inverted – would be older than the rock that lay above it. The limestone was therefore a marker that enabled Sedgwick to distinguish older and younger layers.[7] Moreover, the Bala Limestone was the first band of rock he had encountered in north Wales that contained a significant fossil record – a swirling assortment of trilobites, brachiopods and crinoids – which meant he might be able to match them to similar stones elsewhere.[8] Finally, the dip of the strata seemed to confirm Sedgwick's findings of the year before. If the more easterly band of limestone at Milltir Cerrig dipped to the north-west while the band nearer Lake Bala dipped in the opposite direction, then a cross-section through the area would show them meeting in a bowl or 'syncline' just to the west of the Berwyn peaks.[9] They would, in short, appear to be a continuation of the folds he had plotted running through north Wales the year before.

Many of the outcrops of rock are little more than indistinct patches, virtually indistinguishable from the surrounding mud and clay and buried under dense vegetation.[10] Yet at each site Sedgwick would have had to pause to check its composition and measure its strike and dip. Joseph Beete Jukes, a young Cambridge graduate who had studied under him, was later employed by the British Geological Survey to check his observations. In a series of letters Beete Jukes describes the enormous difficulties he faced tracing the limestone bands, at one point breaking into doggerel to make his point:

> I had a long ramble over Moel Mugynau
> and around about it
> and down the valley we took our way ...
> where Aronfechan's waters play.
> But devil a limestone therein lay.[11]

It would take Beete Jukes more than six months to map them. His letters talk of outcrops 'concealed by the black bogs of these barren moors', and he wrote in despair after one particularly demoralising expedition that 'from the difficult nature of the country I expect it will be almost impossible' to link the outcrops in a coherent pattern. A few weeks later he reported on a colleague who was forced to climb a sharp precipice in pursuit of evidence: 'I did not dare to watch,' he wrote, but 'I hope it will please heaven that the Bala Limestone will never get into such a place again or I will throw my hammer at it, turn round and have nothing more to say to it.'[12]

His colleague Andrew Ramsay, later head of the Geological Survey, encountered similar problems:

> When the geologist first attacks the country he is everywhere met by conflicting dips, strikes and unexpected patches of limestone and ash scattered hither and thither in apparently inextricable confusion, conveying the impression that not one or two but many short beds of limestone and ashes are interstratified with the other rocks; and it was only by the expenditure of much labour and skill that Mr Jukes was able to dissect the main features of this intricate puzzle and prove that all the limestone beds are parts of one and the same bed and that there are only two lines of ashy strata.[13]

As a well-informed local geologist wrote to me in a letter, 'how on earth Sedgwick traced [the limestone] south I cannot imagine. I would not attempt to follow it even with the map.'[14]

One day I tried to walk in their footsteps, tracing a small fragment of the western band of limestone as it threaded its way across the wild and empty slopes of the Berwyn Hills south of

Lake Bala. Lapwings rolled and tumbled in the deep silence, and clouds sat on the tops of the higher hills like volcano smoke. Tangled fences of barbed wire blocked the lower slopes. It was exhausting work. Small outcrops of rock peeped out from the bog and moor, often a mile or more apart. Sometimes the outcrops were plainly limestone, at other times more obviously slate or sandstone. At a spectacular series of precipitous crags known as Craig-yr-ogof, where climbers perch on tiny ledges high above the surrounding country, there was a large limestone outcrop. But further south the band of limestone disappeared under miles of bog and moorland, surfacing only at the bottom of a hill near the tiny settlement of Blaen-y-pennant, a scattering of houses running along the floor of a narrow valley carved out by a fast-flowing stream. Further south still, the limestone vanishes again, before making a final appearance in the shallow stony bed of the River Dyfi, near the remains of a derelict flannel mill in the hamlet of Abercywarch.

I have no idea how Sedgwick pulled off his feat. Did he stumble upon his first sightings of the limestone at Milltir Cerrig or was he tipped off by local geologists? Was he shown the remote and obscure limestone quarry at Gelli-grin or did he find it for himself? We know that he frequently found locals to guide him, but they get no mention, and we know nothing about them.[15]

We can be reasonably certain, however, that unlike Murchison's discovery of the junction of the Old Red Sandstone and Greywacke in the Wye Valley the previous year, this was a genuine eureka moment. Sedgwick knew now how he would spend the rest of the summer. He needed to firm up his speculations about the structure of the rocks to the east and west of the Berwyns and, with luck, confirm that they really did form an older sequence than Murchison's succession of Greywacke rocks further south.

*

Murchison, meanwhile, criss-crossed the hills around Wenlock Edge, the magnificent limestone scarp that dominates southern Shropshire. Years of fox-hunting had given him what was known as an 'eye for the country': a blend of knowledge and experience that let him rapidly grasp the major topographical features of an area,[16] while his training in military draughtsmanship enabled him to quickly transfer this information onto paper. As a result, in a remarkably short period of time he was able draw a series of cross-sections through the hills of the region, showing that the alternating bands of limestone and shale he had noted the year before were literally etched into the contours of the landscape: the land rising and falling in a regular pattern of 'scarps' and 'dales' as it crossed ridges of hard limestone and valleys of softer shale.

Today you can follow in Murchison's footsteps and trace this so-called 'ridge and vale' topography by driving north out of Ludlow along the wide, shallow valley of Corve Dale, carved from softer shales. To the west, tiny lanes climb up the valley slope and onto the harder limestone of the Aymestrey Escarpment and View Edge, then continue to rise and fall in a series of switchbacks as they drop down into the softer shales of Hope Dale, rise again as they hit another band of harder limestone at Wenlock Edge and then fall once more through steeply wooded slopes into the shales of Ape Dale.

If you pick through the rocks of a derelict quarry on Wenlock Edge, the floor overrun by vigorously sprouting saplings and riotous coils of ivy, it takes a considerable feat of imagination to grasp the age and history of these rocks, so ordinary in appearance. Today it is estimated they are more than 400 million years old, a timeframe that renders human existence almost irrelevant. They were deposited in circumstances that, as you look

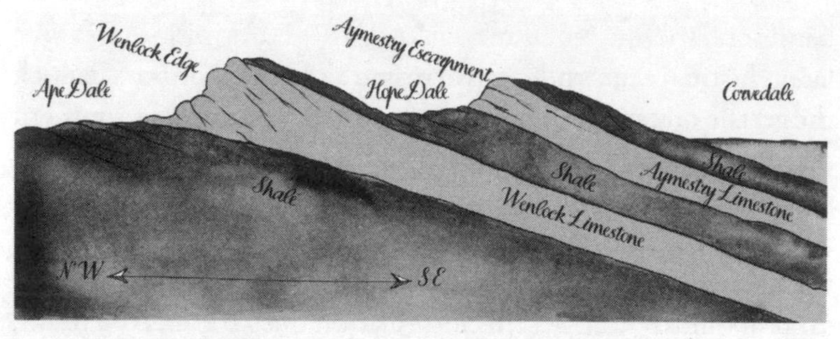

3.1 A simplified section through the ridge-and-vale countryside of southern Shropshire. Murchison would get it broadly right in 1832, but was able to fill in more detail in the following years.

out across a tranquil landscape of fields and farms, are equally difficult to imagine. This was a period when the entire region formed the coastal shelf of a long-vanished land mass to the east. Warm shallow water stretched as far as the eye could see: trilobites such as *Calymene blumenbachii*, the 'Dudley Bug', drifted on ocean currents, while corals spread across a sunlit ocean floor, laying down what we now recognise as the Wenlock Limestone: possibly the first coral reef on Earth.

Over millions of years movements in the Earth's crust erected mountains to the east, and as these eroded, fast-flowing rivers carried mud and silt to the sea, smothering the coral and turning the seabed to sludge. New species of trilobite flourished in the muddy water. This would become the first band of mud and shale, found today in places like Hope Dale. As the mountains to the east wore down, the erosion decreased, the waters cleared and the coral returned. Colonies of the brachiopod Pentamerus, anchored to the sea floor by fleshy stems, gently swayed in the warm shallow currents. Gradually a second band of sediment was laid down: this would become the Ludlow or Aymestry Limestone.[17] And so the pattern continued, with alternating bands of hard limestone and softer shale. Many millions of years later, further movements in the Earth's crust lifted and tilted these rocks to the south-east, and the alternating bands of hard and soft rock were exposed to erosion.

The writer Robert Macfarlane memorably describes geology as equipping us with spectacles that allow us to look back in time to when 'rocks liquefied and seas petrified, where granite slops about like porridge, basalt bubbles like stew and the layers of limestone are folded as easily as blankets'.[18] 'A country is not just a jumble of hills and valleys; the features have a plan, a system underlying their distribution,' wrote the British geologist Arthur Truman equally perceptively in 1937, as the clouds of war closed over Europe. 'The geologist acquires an eye for country and an understanding of nature not excelled by that of the artist or the poet.'[19] To Murchison, for whom most of this extraordinary story was unknown, the strata around Wenlock Edge must all the same have brought a moment of exhilaration. The year before, the idea that the 'upper' Greywacke consisted of a sequence of strata, which could be identified by their individual fossil fingerprints, was entirely new. Now he wrote in his journal that he could confirm this pattern of rocks, as 'numerous details in Wenlock Edge and the Ludlow formations were worked out'. But there was more to come.

The previous winter Murchison had studied maps and notes compiled by earlier mineralogists who had surveyed the region's rocks. He was particularly interested in the work of a man called Arthur Aikin, a one-time Nonconformist preacher from Shrewsbury and founding member of the Geological Society.[20] Twenty years earlier Aikin had made extensive notes on southern Shropshire, including a cross-section through Wenlock Edge, which showed the familiar bands of limestone and shale but also, peeking out from just below them to the west, an entirely different band of rock that he described as 'reddish-brown Flagstone and Sandstone for building'. Leading members of the Geological Society had urged Aikin to publish his findings, but after a six-year struggle to raise money for a publisher, he had given up and printed them himself as a limited-edition pamphlet.[21] A copy found its way to Murchison, who now stealthily began to pick up Aikin's trail of reddish-brown slates and sandstones.

They formed colourful beds of green, purple and red sandstones interwoven with discontinuous threads of limestone. They could also be seen in the walls of houses in local villages, Longville-in-the-Dale, Hope Bowdler and others with equally charming names. It was clear to Murchison that they represented yet another distinct stratum in the Greywacke: what he would come to call 'shelly sandstone'. His stratigraphic table of the 'upper' Greywacke was growing: to the bands of limestone and shale he could now add a new zone, a thick layer of sandstone sitting directly below them. Murchison followed the beds west across southern Shropshire and into Wales, probably with the help of a new inch-to-the-mile map compiled by the Board of Ordnance, the forerunner of the Ordnance Survey, which contained brief descriptions of the region's rocks around the margins.[22]

Once again he also drew on his extensive network of contacts, receiving 'much hospitable treatment from the lords and gentlemen of the region', some of whom were old friends, 'who still enjoyed a quiet joke over the enthusiasm with which he now hunted "Grauwacke" instead of foxes'.[23] He became a regular visitor to Powis Castle, a huge fortified medieval country house standing prominently on an outcrop of red sandstone and Greywacke Limestone, where the Tory peer, Edward Herbert, Earl of Powis, lived with his wife, multiple children and an extensive library.

Murchison paused, too, at Oakley Park, a large house and grounds just north of Ludlow owned by Robert Henry Clive, a descendant of Clive of India, whom Murchison had met during his brief incursion into the Peninsular War. Gradually he moved south; by the end of July, Roderick and Charlotte were back in the Radnorshire home of his friend and host of the previous year, Sir Thomas Frankland Lewis. Charlotte, an attractive and intelligent woman, was widely regarded as good company. There geology blended with merry-making.

'The party divided between croquet and archery,' ran a contemporary account. 'The targets were pitched in the long green narrow meadow which ran down to the river ... and ... it was a pretty sight to see the groups of ladies with their fresh light dresses moving up and down ... At 6 tea, coffee, cider cup etc was laid out and when three dozen arrows had been shot we left off shooting and went to tea.'[24] On wet days guests would retire to the great hall to play battledore and shuttlecock (a forerunner of badminton).[25] Murchison would later conclude that 'good living in an aristocratical mansion is hostile to geological research.'[26]

The Murchisons continued south through the Welsh hills towards the Brecon Beacons, tracking the band of 'shelly sandstone' through a landscape of open skies and wide valleys, flocks of sheep 'hanging on their green slopes'.[27] In quarries and road banks along the way, Murchison began to find a narrow band of dark slate frequently associated with the remains of a large trilobite called *Asaphus buchii*, easily distinguished from its cousins by its size and rounded tail. Sometimes this band was little more than a mere streak, but in all cases it sat at the bottom of the colourful shelly sandstone. Murchison began referring to it as 'Builth and Llandeilo Flags', after some of the locations in which he had found it; or, more generally, as 'Black Trilobite Flagstones'. It seemed to warrant further investigation.

But something inside him rebelled. As he descended the hills to the small town of Brecon, Murchison collapsed with exhaustion. In the isolation of the Welsh hills he appears to have run short of the drugs that he used to maintain his relentless pace of work. 'He depends on them for his existence to frightful extent,' his friend and collaborator Charles Lyell had written in a letter some years earlier. 'On one occasion we were on an expedition together and as a stronger dose was necessary than he had with him, I was not a little alarmed at finding there was no pharmacy in the place.'[28] The drug was almost certainly laudanum, a tincture of opium mixed with wine or water, so commonplace a painkiller that nobody thought twice about taking it. Whether this was really the cause of Murchison's problems we may never know, but the party was forced to retreat to the comfort of the Castle Hotel in Brecon, where Murchison and his wife spent three days resting and gazing out at the Old Red Sandstone of the Brecon Beacons. '1st Big Sunday to recoup, journalize and recover – overdid my work – brought on ?? fever,' Murchison noted in his journal. 'The Surgeon Watkins a rough practitioner but the chemist old Vaughan a kind and intelligent man.'[29]

By a strange stroke of fate, Sedgwick was meanwhile in a similar position. After plotting the course of the Bala Limestone and speculating on the structure of the surrounding rocks, he had set off on a series of mammoth traverses across a landscape of stony paths, peat bogs and drystone walls, to check his observations. Once again he had been confronted by 'vast unstratified masses of porphyry',[30] with none of the clear lines of strata that Murchison was finding further south. By August he had reached the Welsh holiday resort of Barmouth, sandwiched between sea and mountains. Carriages clustered along the flat, sandy sea front and families were enjoying the new fashion for sea bathing. Here, 'burnt as brown as a packsaddle and a little thin from excessive fatigue',[31] Sedgwick also collapsed with exhaustion, complaining of a feverish cold and a sore throat. 'I have toiled like a slave, and have made myself ill, so that I am now almost confined to the house,' he wrote to Murchison.[32]

To add to his woes, without colleagues or a wife to accompany him, Sedgwick was finding the trip increasingly lonely. He was naturally garrulous and sociable. In Cambridge his rooms bustled with friends who were glad 'to enjoy his original conversation and not seldom his extravagant fun'.[33] In Wales he was cut off from such companionship. 'Now and then I have stumbled on a struggling Cantab [former Cambridge student], with whom occasionally I also contrived to spend the evening. These are, however, rare occasions,'[34] he grumbled, of the long nights spent alone among stone cottages and half-starved cattle, where the only language spoken was Welsh and his only companionship was an oil lamp and smouldering turf fire.[35] His field notes for this period contain touching attempts to learn Welsh phrases, such as 'Good Morning', 'Good night', 'How do you do' and 'Thank you, can you give me some milk and have you any mutton/beef/trout/fish?', but he was well aware that he was excluded from most passing conversation. 'I like to talk to country people, and to see their humours, but from this I am shut out among these children of Caractacus. This it is which has made my solitude doubly solitary,' he wrote to a friend.[36]

In the absence of company that summer, Sedgwick followed the customs of the time and passed the hours drinking. As a typical contemporary noted, 'I drank 3 pints (of port wine) every day besides 1/3 pint of brandy.'[37] The consequences of such constant indulgence were inevitable, but poorly understood: Sedgwick, like many of his contemporaries, suffered from that 'hissing hot fit' of nineteenth-century man, that 'well-known tamer of the most forward spirits' – gout.[38] 'Since we parted I have had a short visit from my old and very unwelcome acquaintance which has affected my arterial system and produced a throbbing in my head which I must get rid of before I can fairly set to,' he wrote to a friend in Cambridge.[39] 'My ailments are a flea-bite; but to have a flea always biting, and to be always conscious of the bite, would be bad enough in all conscience. And such is my case.'[40] Men 'drank freely', recorded one observer, 'and accepted gout and apoplexy as things that ran mysteriously in respectable families.'[41]

Finally after several days the pain subsided and Sedgwick rose from his sickbed and returned to the hills, covering up to 40 miles a day. He zigzagged south through the uninhabited moorlands of central Wales, the 'biggest, brutalest, brownest, boggiest moors in the country', according to Beete Jukes,[42] before returning north to explore the eastern flanks of the Berwyns – a round trip on horseback and by carriage of about 100 miles. It was 'the severest summer's task of my geological life', he would later reflect. But he had amassed a huge amount of data in support of his ideas.

West of the Berwyns, as he had predicted, the strata rose towards the Arenigs and the so-called 'Merionethshire saddle': and to the east he had identified a similar dome running down through the Welsh border country, which suggested that the Bala Limestone would re-emerge somewhere near the small village of Meifod on the eastern flanks of the Berwyns. This last conclusion was particularly important. It suggested that the strata at Meifod dipped to the south-east and therefore probably dived under the rocks lying to their south and east: an indication that the strata that Sedgwick was exploring were older than the border ones Murchison had been investigating to the south. 'The beds ... roll over to the SE so as to bring in the *newer* rocks

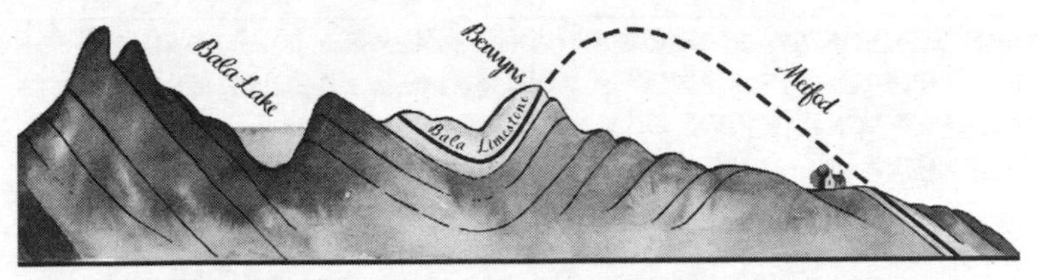

3.2 Sedgwick's cross-section through the Berwyns. It confirmed that to the west the rocks rose towards the 'Merionethshire saddle' and Harlech Dome, and to the east formed the Montgomeryshire dome. As a result, the Bala Limestone was likely to reappear near the small village of Meifod, suggesting that the rocks of north Wales were diving under younger rocks to the south-east.

... which form the base of the system in which you are working,' he wrote to Murchison (emphasis added).[43]

That autumn, in lectures to the Cambridge Philosophical Society, Sedgwick drew up his first rough-and-ready stratigraphic column of the rocks of north Wales. It consisted of four broad zones. With the exception of the Bala Limestone, they were as yet ill-defined and lacked distinctive fossil markers. Indeed, he would describe the upper band of his 'Berwyn Blue Slates' as 'a great obscure group of which I don't exactly know the end or side', and it was distinguishable from the lower band of 'Snowdonia Slates' only by their position relative to the Bala Limestone. This made it difficult to connect them with rocks elsewhere, including those to the south that Murchison was exploring. As beginnings go, it was a tentative one. But it was progress all the same.

By now the Murchisons had returned to London, where Roderick boasted to a friend, 'I have done a fine stroke of work ... and I hope to show you four or five distinct natural fossiliferous formations of great thickness in our neglected grauwacke in which

MURCHISON'S UPPER GREYWACKE SEQUENCE	LUDLOW AND WENLOCK LIMESTONE	The sequence of shale and limestone beds lying just below the Old Red Sandstone. The Wenlock rocks, in particular, were full of fossils including crinoids, brachiopods, corals, graptolites and trilobites.
	SHELLY SANDSTONE	Green, purple and red rocks described as 'shelly sandstone' though they were actually a mixture of sandstone and limestone. They shared a similar 'fossil fingerprint' of trilobites, brachiopods and crinoids.
	LLANDEILO OR BLACK TRILOBITE FLAGS	Dark-coloured flagstones characterised by the fossil remains of the trilobite *Asapus Buchii.*
SEDGWICK'S LOWER GREYWACKE SEQUENCE	BERWYN BLUE SLATES	Rocks lying above the Bala Limestone that covered the higher reaches of the Berwyns and much of South Wales. They contained almost no fossils.
	BALA LIMESTONE	A mix of limestone and shale beds, relatively rich in fossils crinoids, brachiopods and trilobites.
	SNOWDONIA SLATES	Rocks lying below the Bala that made up most of the mountains of north Wales. Like the Berwyn Slates, they appeared to contain very few fossils.
	PRIMARY SLATES	**The original Primary rocks. A confusing category that included the very unslate-like granites and schists of Anglesey and the Llyn Peninsula, as well as the slates of the western slopes of Snowdonia.**

3.3 A stratigraphic representation of Sedgwick and Murchison's first attempt, in 1832, to identify and categorise the Greywacke rocks of Wales and the border country. They agreed the rocks identified by Murchison were probably younger than those identified by Sedgwick and therefore rested on top of them. They also agreed they probably represented two different stages in the Earth's early history: what they now began to cautiously refer to as the Upper and Lower Greywacke.

I have got abundance of fossils – many quite new.'[44] True to his word, he produced his own crude stratigraphic column broken down into three zones of rock: a band of Ludlow and Wenlock Limestone and shale, the colourful band of 'shelly sandstone' and, at the bottom of them all, the tentatively identified bed of Llandeilo or Black Trilobite Flags.

During the winter and spring of 1833 Sedgwick and Murchison compared notes. They agreed that Sedgwick's series of rocks were probably older than Murchison's and that their two stratigraphic columns probably represented two different stages in the Earth's early history: what they now began to refer to cautiously as the Upper and Lower Greywacke. Murchison, with neither a university nor ecclesiastical position to take up his time, and conscious that his reputation stood or fell on his fieldwork, was eager to go public with their findings. His term as president of the Geological Society had another year to run and he planned to use his presidential address to announce the discovery of an Upper and Lower Greywacke. But Sedgwick was more cautious: there was still considerable uncertainty about how and where their two periods dovetailed. It was just about possible that one simply sat on top of the other, but that would have been a remarkable coincidence, given that the division of their two areas of fieldwork had been entirely arbitrary. It was therefore equally possible they were not dealing with two discrete periods and there was an overlap between them. 'I am not so confident,' wrote Sedgwick, 'that I should like to commit myself in any formal précis.'[45]

In the event Murchison swallowed his frustration and confined his presidential address to the so-called Upper Greywacke. The two men agreed to suspend judgement on how their two periods married up – for the time being.

*

As the spring of 1833 turned to summer, Murchison hurried back to Wales to double-check his findings. He was anxious to clarify the exact relationship between the Upper and Lower Greywacke and trace out the geographical boundary between them. He invited Sedgwick to join him, even telling him that the party would include an additional lady who might interest him. But Sedgwick, laid up in Cambridge with another severe attack of gout, was in no mood for female company or another summer in the Welsh hills and replied with strange venom, 'she is a most formidable and cruel tyrant, who has slain tens of thousands without pity. I should not like to be ... a burnt offering', adding for good measure, 'besides, what have I to do with love ... I am wedded to the rocks.'[46]

The Murchisons crossed south Wales to the valley of the River Towy where, in quarries and along the river banks, they picked up further bands of Black Trilobite Flags they had identified the year before and followed them north into Shropshire. They were effectively retracing their steps and checking their fieldwork of the year before. Once again they received considerable assistance from a network of local geologists and 'quarry men'. 'Each succeeding year in which I propagate the principles of our craft and enlist raw recruits in provinces where the sound of the word geology was never heard before,' Murchison wrote patronisingly to a friend, 'I find on revisiting my fields of battle that my aides-de-camp have collected facts.'[47]

In many places he found that the band of Trilobite Flags was missing and there was a gap, or unconformity, in the rock record. In other places the rocks were so shattered and faulted it was impossible to read them with any accuracy. But by the end of the summer Murchison was convinced he could trace a line of crumbling black slate and flagstones running from Llandeilo in the south through to the village of Longnor in central

Shropshire. This clearly marked the border between the Upper and Lower Greywacke.[48] If he was right, he could now divide Wales into two zones: a region of older Greywacke to the north and west and an area of younger rocks to the south and east.

That wasn't all, however. Sifting through the five boxes of fossils he had collected from the Greywacke, Murchison noticed that they consisted entirely of marine animals: a diverse mixture of molluscs, brachiopods, trilobites and corals. There was not a single one of the swamp- or land-based trees and ferns that were characteristic of the coal-bearing Carboniferous rocks. For Murchison, well aware that geology was one of the key sciences driving the Industrial Revolution, this had important implications: large areas of the country could be ruled out at a stroke from the hunt for coal, and he could advise all landowners living on the Greywacke that they would be wasting their money by sinking exploratory pits.

As for the mystery of the Greywacke, as far as Murchison was concerned, it was solved. The boundary between their two systems was clearly marked by the Black Trilobite Flags and he was now proposing a 'big book' on the Upper Greywacke. All that remained to be done was to pick up where he had left off and trace the boundary between the Upper and Lower Greywacke, north from central Shropshire to the Irish Sea. It looked like a relatively simple operation.

Neither he, nor anybody else, could have guessed how incorrect he would turn out to be.

Chapter Four

A NEW HISTORY: THE CAMBRIAN AND SILURIAN, 1834–5

In early June 1834 Sedgwick and Murchison joined the summer flood of tourists heading for Great Malvern to take the spa town's fabled 'water cure': a brutal daily regime of cold-water bathing, exercise and dieting, said to cure everything from fatigue to stomach problems. Their plan was to confirm their findings of the year before and then thrash out the remainder of the boundary between the Upper and Lower Greywacke as it ran north from Shropshire to the Irish Sea. It would be their first joint trip to the Welsh rocks and, if successful, would lay the basis for the first comprehensive geological map of the region.

Both men were keenly aware that the eyes of the geological community were on them. Some weeks earlier the Cambridge philosopher and leading member of the Geological Society, William Whewell, had written to Murchison's wife Charlotte, only half in jest, urging her to 'fall in with them in time to prevent their turning their fratricidal hammers on one another ... which I fear would be the result if they could not agree about the dovetailing of the two portions'.[1]

They left Malvern and made for the Herefordshire market town of Ledbury. At a nearby quarry they observed bands of limestone and shale that marked the top strata of Murchison's Upper Greywacke series. Swinging west through the Brecon

Beacons and south Wales, they traced the line where these upper rocks met the Old Red Sandstone immediately above them. This was the boundary Murchison had logged previously and it gave them an agreed point where the Greywacke met the Secondary rocks. Three days later they reached Llandovery and the rolling pastures of the Towy Valley, where at sites along the river the tilted, vertical strata plainly showed a band of Black Trilobite Flags sitting on top of beds of 'blue calcareous slates' that resembled the blue-grey slates of Sedgwick's Lower Greywacke. They had passed from the top to the bottom of Murchison's Upper Greywacke and located the junction where it passed down into Sedgwick's lower system of rocks.

Murchison would later report that Sedgwick 'endeavoured to pick every hole he could in my arrangement', but 'has confirmed all my views, some of which, from the difficulties which environed me, I was very nervous about'.[2] Perhaps to the surprise of both men, their entirely arbitrary division of the rocks two years earlier appeared to reflect a genuine geological distinction. Murchison's Upper system seemed to sit directly on top of Sedgwick's Lower one.

Their small party turned north, along 'deep narrow lanes and steep plunging hills',[3] following Murchison's travels of the previous summer and tracing intermittent outcrops of the Black Trilobite Flags in river valleys and small quarries. They worked at speed: within a week they had agreed the line of the boundary between the Upper and Lower Greywacke as it ran from south Wales to Shropshire – the very line Murchison had mapped the year before. By the end of June they were at the steep slopes of the Berwyn Hills and the point where their two areas of fieldwork met.

It was here that they hit their first problem. In a quarry near the small village of Meifod, a former religious centre on the southern flanks of the Berwyns, they uncovered a band of sandy limestone and shale that looked remarkably similar to Murchison's 'shelly sandstone'. It was composed of limestone boulders surrounded by a bed of shale and sandstone, and contained an almost identical fossil fingerprint. Yet, to Sedgwick, it looked strikingly like the band of limestone boulders he had identified two years earlier in the quarry at Gelli-grin and assigned to his Bala Limestone strata. Moreover, this was exactly the point where he had predicted that his band of Bala Limestone would reappear as it rolled down the eastern flank of his 'Montgomery anticline'. Both men, it seemed, were laying claim to the same band of rock. Could it be that there was an overlap between their two systems that until now they had completely missed? But in that case the carefully plotted boundary *between* them would make no sense.

Almost immediately, and apparently without serious argument, Sedgwick backed down and amicably assigned the Meifod sandstone to the Upper Greywacke. There is no record of their conversation, so we will never know why he made such a rapid concession; it would mean entirely rethinking his system of the Lower rocks.

With the crisis avoided, however, the two men pushed on into the higher Berwyns and what they assumed would be Sedgwick's older rocks. They appear not to have stopped to explore the band of Bala Limestone near the Milltir Cerrig pass, but instead rode over the bleak tops of the Berwyns and descended the zigzag track down the western flank of the hills to inspect Sedgwick's Bala Limestone at Gelli-grin. Here, under a green canopy of oak, ash and birch, it quickly became apparent that they had hit a second problem. To Murchison, the line of limestone boulders looked extraordinarily similar to the limestone rocks at Meifod that they had just assigned to the Upper

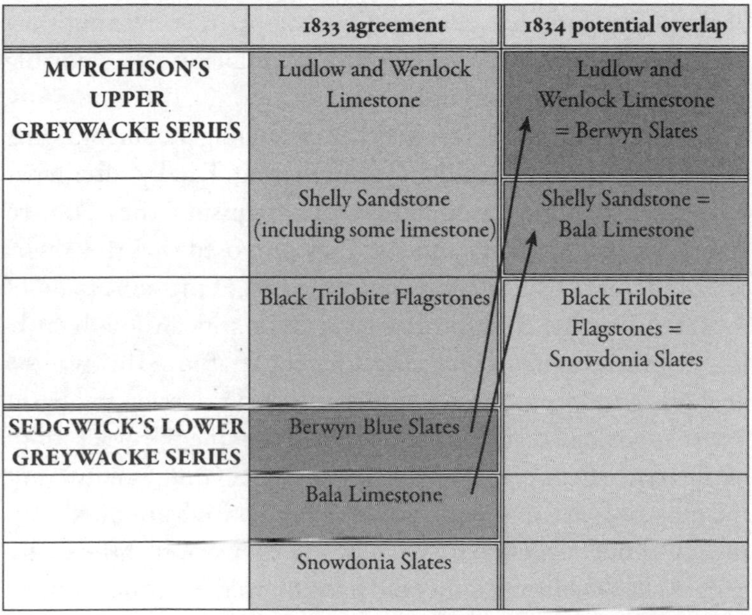

	1833 agreement	1834 potential overlap
MURCHISON'S UPPER GREYWACKE SERIES	Ludlow and Wenlock Limestone	Ludlow and Wenlock Limestone = Berwyn Slates
	Shelly Sandstone (including some limestone)	Shelly Sandstone = Bala Limestone
	Black Trilobite Flagstones	Black Trilobite Flagstones = Snowdonia Slates
SEDGWICK'S LOWER GREYWACKE SERIES	Berwyn Blue Slates	
	Bala Limestone	
	Snowdonia Slates	

4.1 The moment of crisis. If Murchison's Shelly Sandstone and Sedgwick's Bala Limestone were the same, then there was a significant overlap between their two systems and they would need to rethink their stratigraphic column.

Greywacke. There was the same structure of limestone pillars and they contained an almost identical fossil fingerprint: a collection of trilobites, brachiopods and crinoids. Yet both men had assumed they must by now have crossed the boundary out of the Upper Greywacke and were firmly among Sedgwick's older Lower Greywacke. Indeed, neither of them had ever considered the Berwyns to be anything other than Lower Greywacke rock. It was another difficult moment, and I imagine them hammering off the weathered outer surface of the rocks to examine the clean, freshly exposed limestone and wondering what to make of them. As in Meifod, everything they had agreed until now was beginning to look less certain. Over supper that evening they must have wondered whether they were facing the daunting prospect of a major reappraisal.

Over the following days they kept searching the surrounding hills for a solution. Nothing suggested itself. Finally, after what we can only assume was much anxious discussion, they came to – well, let's call it an agreement. They proposed that the 'shelly sandstone' and Meifod limestone were part of the same band of rock, but that the Bala Limestone at Gelli-grin, although strikingly similar, was an older and different stratum. This was, to be frank, a fudge: a compromise that lacked any scientific basis. 'Neither geologist, at this date, could determine the exact point of difference between the two,' wrote a later commentator, but they appear to have agreed 'to respect each other's unwritten territorial rights.'[4] Thanks to this gentlemanly division of the rocks, Sedgwick and Murchison could hang on to their concept of an Upper and Lower system, in the vague belief that somehow the detail could be thrashed out later.

The area north of the Berwyns was Sedgwick's territory, and the job of completing the map of the boundary now fell to him. There was a severe shortage of fossils, including the Black Trilobite Flags that Murchison had used as a marker further south. But Sedgwick, accustomed to working in rocks with a minimal fossil record, was already considering a different approach.

For some years he had been following the ideas of a prominent French geologist called Jean-Baptiste Élie de Beaumont, the engineer-in-chief of France's mines and the man charged by the French government with overseeing the country's first geological map. Élie de Beaumont was a proponent of 'parallelism', a theory based on the idea that the Earth had once been a fiery globe and had been cooling and shrinking ever since. As it did so, the surface would have suffered a series of rapid and dramatic contractions, which pushed up a succession of mountain chains, each characterised by the fact that the mountains produced by a single 'contraction' all ran in the same broad direction across the Earth's surface. From this, Élie de Beaumont was able to surmise that all mountains with the same axis of orientation must belong to the same period in the Earth's history. Furthermore, all mountains with different axes of orientation must have been thrown up by different contractions and therefore belonged to a different period.

It was controversial and speculative, but Sedgwick grabbed at it and, as he now travelled up through north Wales, applied the theory to the hills around him.[5] All the way up through the Berwyns the bands of strata appeared to run in a roughly north–south direction. But as he reached Thomas Telford's 'Irish' turnpike (now the A5), as it cuts through the northern flanks of the hills, the orientation of the strata seemed to change. Instead of running north–south, they appeared to run in an east–west direction, suggesting they were of a different age.

Once again Sedgwick embarked on an epic trek through a wilderness of open moorland. Heather, moss and marram grass stretched to the horizon. Grouse rose in front of him as he struggled from one outcrop to another. Occasionally he encountered small disused quarries, where the rock had been excavated to build a house or field boundary. Battling across the open moorland, he traced 'a very crooked line' line through the hills where the orientation of the strata changed. It ran north along the eastern flank of the Berwyns until, somewhere near the slate-mining village of Glyn Ceiriog, it turned sharply west through the hills of the Arenigs; then north again in the valley of the Conwy River, until it hit the sea near Colwyn Bay.

Somewhere along this arduous trek Sedgwick appears to have also made an important inference. Since the Berwyn Hills were considered to 'belong' to his Lower Greywacke and were therefore among the oldest hills in Wales, the rocks to the north, with their differently orientated strata, must belong to a younger period and form part of Murchison's Upper Greywacke. In one bold swoop he had found his boundary between the Upper and Lower Greywacke.

Sedgwick drew a map of Wales to illustrate this. Murchison was triumphant. After three years of hard – even heroic – fieldwork, the undifferentiated mass of the Greywacke, the 'Red and Dunstone' and 'Killars and Slate' of William Smith's original geological map of Britain, had been broken down into discrete categories and mapped. The so-called Transition rocks of Wales had been divided into an Upper and Lower system and, despite certain local difficulties, the boundary between the two appeared to be clearly marked by the layer of Black Trilobite Flagstones in the south and Élie de Beaumont's 'axis of elevation' in the north. Murchison wrote to William Whewell, the man who at the start of the summer had worried that the two geologists might fall out over the boundary between their two systems, that they had 'parted company on the most friendly terms ... having dovetailed our respective upper and under works most satisfactorily for both of us'.[6]

There was only one issue remaining to be resolved: the point at which they had chosen to draw the boundary between the Upper and Lower series was entirely arbitrary and simply reflected the different areas of the country they had selected to explore. Whether the Shelly Sandstone and Black Trilobite Flags fitted into an Upper or Lower system was purely a matter of where you chose to draw the line. There appeared to be nothing that gave either the Upper or Lower group a particular internal logic. They were simply a descriptive convenience. Murchison had occasionally suggested that the Upper Greywacke was characterised by a unique fossil fingerprint, but the similarity of the fossils in his Shelly Sandstone and Sedgwick's Bala Limestone made this idea hard to maintain.

It was now that Élie de Beaumont made a second contribution. He had heard about the work they were doing in Wales, and during the winter of 1834–5 produced an imaginary sketch of the Transition rocks and the relationship between Sedgwick's and Murchison's systems. In doing so, he had marked an 'unconformity' or break in the rock record at the bottom of the Upper Greywacke where it joined the Lower Greywacke, and another at the top of it where it met the Old Red Sandstone of the Secondary Rocks. Élie de Beaumont's sketch was merely an illustration. It was not based on any systematic geological research. But when Murchison saw the diagram he realised it was just what he needed.

It implied that at both the top and the bottom of the Upper Greywacke major changes in conditions on the Earth's surface had caused a break in the deposition of the rocks. That could have been a fall in sea levels triggered by climate change, or a rise in the land following disturbances deep in the Earth's surface: to Murchison, it didn't matter. What it suggested was that the Upper Greywacke was much more than simply a random grouping of strata; it represented a distinct and coherent period in the Earth's history, bookended by radical change. What had started out, two years earlier, as an entirely arbitrary division of the rocks had become something much more significant: a distinct geological period that was recognisably different from those that came before and after it.

That December, at a meeting of the Geological Society in London, Murchison announced he would be dropping the name 'Upper Greywacke' for something more appropriate to its new significance. He chose the name 'Silurian', after a Celtic tribe that had once lived on the Welsh borders. 'I had seen,' he later explained, 'that all geological names founded on mineral or fossiliferous characters had failed to satisfy, and that fanciful Greek names were still worse. Hence it seemed to me that a well-sounding geographical term, taken from the very region wherein the classification had been elaborated, and where every one might go and see the truthfulness of it, was the best.'[7] It no doubt helped that the Silures had excelled in military manoeuvres and warfare; indeed, some years later Murchison would praise their chieftain, Caractacus, as a 'wily' and 'gallant leader'. But this was more than a simple matter of finding an appropriate name.

Since the demise of his military career, Murchison had been looking for a purpose in life: a project he could make his own that would lend him standing and respect in society. Finally he had found it. In devising a name for the Upper Greywacke rocks, he was also laying claim to them and all they represented. He was now no longer merely 'invading' the rocks, he was annexing them. His letters and writing contained increasing references to 'my' rocks and 'my' Greywacke. Furthermore he now felt sufficiently confident of his findings to consider publishing them as a book. He hoped that Sedgwick, long committed to writing a book on the Transition rocks with William Conybeare, would collaborate and the three men would be able to publish a complete account of the entire series of Transition rocks. But that winter Sedgwick had other preoccupations that pushed the Greywacke to the back of his mind.

In November 1834 he was offered a part-time job at the wealthy, well-endowed cathedral in Norwich. This would dramatically improve his finances. For the past fifteen years he had struggled to live on his meagre income from the Woodwardian professorship and his part-time job at the village church in Shudy Camps. The Norwich post, known as a prebendaryship (in effect, a part-time canon's job), offered an additional salary of £600 – six times his Woodwardian salary, and the equivalent of around £35,000 today – for a mere two months' work a year. The two-month tour of duty, furthermore, would fall during the winter months of December and January, when geological fieldwork was at its least attractive.

Sedgwick seized the opportunity 'without a moment's hesitation'.[8] He was forty-nine years old and, for the first time in his life, would be able to live comfortably on what he earned. But it meant that during the winter of 1834–5, as Murchison pushed ahead with his 'big book' on the Silurian System, Sedgwick was in Norwich, distracted by what he described as an incessant daily round of public duties, starting with an 'interminable' early-morning service in the 'cold empty cathedral' and continuing through to midday when there was a second long service. It left him, he told Murchison, 'dog tired and unfit for work'.[9]

Back in Cambridge in early February, life was no quieter. 'Since my return I have been almost driven off my feet,' Sedgwick complained. 'Lectures, college business, arrears of correspondence, disagreeable domestic news ... and more than all together the oppressive consciousness of having more work before me than I have any chance to get through.'[10] The sense of overwork was exacerbated by regular crippling bouts of gout and rheumatic fever, which left him incapable of getting out of bed for weeks on end. In an effort to treat them, he devoted yet more time to experimenting with fashionable medical remedies such as 'dephlogistication', a procedure that involved drinking large quantities of water to flush out 'phlogiston' – a theoretical substance believed to be associated with fire and, by extension, with fevers and high temperatures.

To his colleagues it was also becoming clear that Sedgwick, despite his other commitments, 'had no intellectual self-control', as one put it. 'He could never shut his eyes and ears to what was going on around him' and, as a result, the geology of the Greywacke was 'laid aside for long intervals, because he ... allowed himself to be carried away by something foreign to what ought to have been the real purpose of his life.'[11]

One of these was his post as a senior university proctor, a sort of university policeman responsible for supervising the physical and moral conduct not only of the student body, but of the local population as well. 'Behold me now!' he wrote to Murchison, 'in a new character, strutting about and looking dignified, with a cap, gown, cassock, and a huge pair of bands – the terror of all academical evil-doers – in short, a perfect moral scavenger.'[12] The scavenging could last well into the night. 'At 11 this night I was called out of the Chequers [pub] by Webber to go with him and quell a riot in George Lane,' runs a typical entry in a fellow proctor's diary. Later, on returning home, they encountered a 'common strumpet' and 'if it was not for me she would have been sent to the Bridewell. It was one o'clock before I got to bed.'[13] 'Women of the night', real or imagined, became one of Sedgwick's more improbable preoccupations. 'No wandering woman was safe with the Revd Sedgwick on the prowl,' according to another account. 'His object was to eradicate' and 'flush out those infernal females'.[14]

Perhaps it was. Several years earlier in Shrewsbury, when Sedgwick had found a rapport of sorts with Darwin's three unmarried sisters, the young Darwin had cracked jokes about the pending announcement of a 'Mrs Sedgwick'. Three years later friends were still speculating. 'Whether he fell in love with some of the Salopian lasses or not is in his own breast; but I can assure you that a whole houseful of them are deeply smitten,' Murchison wrote teasingly to a friend after the two of them had visited the Shropshire town. 'When we parted ... it was found he had left that beautiful brown coat of his in the very house where all these sirens were, so I left him posting back to recover the old garment, and perhaps to leave his heart.'[15]

It wasn't to be. Friends remarked that 'he would be much happier and would eventually do much more for geology' if he were to marry,[16] but Sedgwick appears to have been simultaneously attracted to, and alarmed by, female company. Perhaps to avoid the 'threat' posed by potentially eligible women, he devoted many hours, when not dealing with ailments or prowling the streets of Cambridge, to maintaining a regular correspondence with the wives and daughters of friends and with his several nieces.[17] 'Have I not sent you a long gossiping letter,' he cheerily signed off a despatch to one.[18] 'I only wish you could come with your Mamma and Annie ... and encourage me with your kind looks and then come home and help me drink tea,' he wrote to another.[19]

Given these demanding rounds of sickness, self-medication, correspondence and moral scavenging, it was not until March 1835 that Sedgwick finally got round to writing up some of the findings from the Welsh tour.

The most significant was a paper dealing with a phenomenon that had caused considerable confusion in areas like north Wales, where geologists had noticed the tumbling cliffs of slate were bisected by thousands of closely spaced, parallel lines of weakness, where the rock naturally split into thin sheets. They looked broadly similar to the strata that were characteristic of sedimentary rock, but bore no relation to the orientation of the bands of strata in neighbouring rocks and, in many cases, overlaid or obscured them. What, geologists wondered, were they? And what had caused them?

In his new paper, catchily titled 'Remarks on the Structure of large Mineral Masses, and especially on the Chemical Changes produced in the Aggregation of Stratified Rocks during different Periods after their Deposition', Sedgwick explained that their superficial resemblance to strata was entirely misleading. They were the result not of the process of deposition, but of extreme heat and pressure in the Earth's crust, which had altered the composition of already existing beds of sand and mudstone. This metamorphism had caused the grains in the rock, which usually form a haphazard matrix that 'locks' the rock together, to line up in neat parallel lines or planes of weakness, where the rock will naturally splinter. To distinguish them from beds of strata, Sedgwick called them (quite innocently) lines of 'slatey cleavage'.[20] This was an important advance in the understanding of the rocks of north Wales, but added only marginally to Sedgwick's understanding of the outlines of the Lower Greywacke.

Friends looked on in frustration. 'S asked me to walk home with him,' noted Charles Lyell that spring. Sedgwick wished to write something, but 'I found a gloom upon him unusual and marked'.[21] Some questioned his commitment to the collaboration. I 'feel sure the work won't be done in a year, nor perhaps two,' continued Lyell. 'He has not the application necessary to make his splendid abilities tell ... The fact is that to become great in science a man must be nearly as devoted as a lawyer, and must have more than mere talent.'[22]

Murchison, still in a hurry to establish himself in London's scientific and cultural circles, was also beginning to find the collaboration frustrating. Until now he had looked to Sedgwick for his geological expertise; now the partnership was causing tiresome delays.[23] In an article in the highly respected *Philosophical Magazine*, a well-established scientific journal of the time, he reiterated his outline of the Upper Greywacke and concluded, 'My friend Prof S will doubtless soon disperse the obscurity which hangs over ... [the Lower Greywacke] rocks.'[24] But soon afterwards, desperate to pin the whole thing down, he complained, 'I can obtain no opinion from him nor will he stick to it without you force him.'[25]

Finally in the summer of 1835, after nearly a year of continuous pressure from Murchison, the two men presented a joint paper to the annual meeting of the British Association for the Advancement of Science, held in Dublin that year. 'The geologists were allotted one of the largest meeting rooms, and as expected their Section was one of the most popular,' writes the historian of geology Martin Rudwick. 'Murchison and Sedgwick held their joint paper in reserve until the last day so that it would come as the climax of the week's work.'[26] It was here that they publicly announced the division of the Greywacke into two distinct geological periods: Murchison's Upper Greywacke or Silurian Period, and Sedgwick's Lower Greywacke or Cambrian Period, named after the Roman word for Wales that, after some deliberation, he seems to have plucked out of the air.[27]

The paper also included a new and more comprehensive stratigraphic table that combined their two existing tables, with some of the strata renamed to make them more distinctive. Murchison's Ludlow and Wenlock Limestones became two separate categories: the 'Shelly Sandstone' became the more memorable 'Caradoc Sandstone', after Caer Caradoc, the hill rising above the once-thriving health resort of Church Stretton, where Murchison had found it in quarries; and the Black Trilobite Flags became the 'Llandeilo Flags', after the small Welsh town where

Murchison had first encountered them. A geological 'terra incognita', Sedgwick told a later meeting, had become a 'known country which, though just discovered, already exhibits every symptom of regularity and improvement'.[28]

UPPER SILURIAN	Ludlow Rocks
	Wenlock Limestone
LOWER SILURIAN	Caradoc Sandstones
	Llandeilo Flags
UPPER CAMBRIAN	Higher beds of Berwyns/south Wales slates
	Bala Limestone
MIDDLE CAMBRIAN	Higher mountains of Caernarvonshire and Merionethshire, including Snowdonia
LOWER CAMBRIAN	The mica schists of Anglesey and south Caernarvonshire

4.2 Murchison and Sedgwick's agreed classification of the Greywacke strata, as announced at the Dublin meeting of the British Association for the Advancement of Science in 1835.[29]

The paper was hailed as a major success. Sedgwick told his brother John that it received 'such a reception as has eclipsed the remembrance of all former meetings'.[30] The new categories had rapid and important consequences. If the Cambrian and Silurian really did represent distinct geological periods, then it should be possible to identify them not just in Britain, but across the world. Almost immediately, geologists in continental Europe began to apply the new categories to rocks from Scandinavia to Spain. One small district of Wales had become the template, or 'type section', for the rest of the world. William Buckland, Sedgwick's opposite number at Oxford, applied them to rocks that he was exploring in the French Ardennes. Élie de Beaumont began using them for rocks across the rest of France. The mapmaker and fellow member of the Geological Society, George Greenough, took the idea to a meeting in Germany. Geologists in the United States began to amend their ideas to fit the new categories.

Human understanding of the Earth's history had pushed further into the recesses of time than ever before. Sedgwick and Murchison started to appear as quintessential Victorian heroes – their long, uncomfortable and, in some eyes, intrepid treks across the Welsh hills contributing to the image of geology as a glamorous and 'romantic' science.

How, though, to cement their triumph? If the two systems were to have universal applicability, it was important that they could be clearly identified by fellow geologists across the world. For Murchison, the best way to achieve this was to build on the work of Thomas Lewis and establish a clear fossil record for each of his Silurian strata: what he was now calling a 'fossil-based stratigraphy' for the Greywacke. To this end, his paper on the Silurian rocks included a detailed account of the unique fossil fingerprint associated with each of the newly named strata. This was particularly useful for correlating beds of Caradoc Sandstones that could, superficially, look very different.

For Sedgwick the task was more difficult. He had no doubt about the importance of fossils; indeed, his job as Woodwardian professor at Cambridge had originally been entitled 'Professor of Fossils' and he continued each year to buy and collect specimens for the Woodwardian Museum. But he had spent much of his time as a geologist exploring upland areas like north Wales, Cumbria and the Lake District, where fossils were notoriously hard to find, and he had grown accustomed to identifying and correlating rock strata by their rock type and position. This was the approach that he had used to draw his cross-sections through north Wales in 1832, and which he had employed the previous year to locate the boundary between the Upper and Lower Greywacke with the help of Élie de Beaumont's parallelism. In fact, with the exception of the Bala Limestone, he would have been hard pressed to draw up any kind of fossil fingerprint for the rocks of the Cambrian Period.

In the summer of 1835 this difference of emphasis remained relatively insignificant. Murchison drew on the structure and rock type to correlate strata when it suited him, and Sedgwick continued to collect fossils whenever he could. But as geologists across the world began to look for examples of Cambrian and Silurian rocks, the idea of a simple fossil-based science of stratigraphy became increasingly attractive. It appeared to offer a clear and objective means of classifying strata, which overcame the vagaries of appearance or the need to understand a region's geological structure. It was 'as if the mighty Maker of all things', wrote one contemporary, 'had stamped each age of life with a seal that would distinguish it'.[31] Murchison, in particular, began to press with increasing insistence for a science of stratigraphy based exclusively on the fossil record. The unique fossil fingerprint associated with each of his Silurian strata seemed to confirm his conviction that each band of strata represented a unique moment in the history of life on Earth. Sedgwick's insistence on the fundamental importance of the structure of the rocks was starting to look increasingly unfashionable.

On the other hand, the apparent lack of a fossil record in the Cambrian rocks raised an intriguing possibility. Could it be that there was a point, somewhere between the Cambrian and Silurian systems, when life itself had begun? Indeed, would the first signs of life be found somewhere along the Cambrian–Silurian boundary? This was the question that turned a relatively obscure debate into a subject of wider interest. And perhaps it was this sense of a waiting audience that encouraged Murchison finally to issue a prospectus for his book on the Silurian rocks.

In a fit of false modesty, he told colleagues, 'his friend Mr Frankland Lewis had suggested that he should not be content with the limited circle of readers which perused the ponderous Transactions of the Geological Society, but should appeal to a wider public, and elaborate into a separate volume his researches among the old rocks of the English and Welsh border-land'.[32] We can be sure Murchison was more than happy to oblige. By November he had attracted some eighty subscribers, including Frankland Lewis and Edward Herbert, the Earl of Powis, each of whom pledged the equivalent of several hundred pounds in today's money towards the publication.

But even as Murchison was gathering subscribers, the integrity of his Silurian System, never completely buttoned down, was facing a major challenge from a completely unexpected quarter.

Chapter Five

THE DEVONIAN CONTROVERSY, 1835–40

Henry De la Beche was an enthusiastic young geologist in the west of England. He had much in common with Murchison. The son of a wealthy family of West Indian sugar-plantation owners with a 4,500-acre estate in Jamaica, he had been sent in 1810, at the age of fourteen, to the country's leading military training college for officers in Great Marlow, the same establishment attended by the young Murchison several years earlier.[1] But the teenage De la Beche had been a rebel and a year later he was expelled for encouraging 'a dangerous spirit of Jacobinism' among the cadets.[2] The family moved to the increasingly fashionable seaside resort of Lyme Regis on Dorset's wild and fossil-rich Jurassic coast and a footloose and wealthy De la Beche took to sailing, chasing girls and collecting fossils, often in the company of a young local woman called Mary Anning, who was later to become one of the most prominent fossil-collectors in Britain.[3]

An anonymous satirical poem about life in Lyme Regis in the early 1800s describes the activities of a 'Sir Fopling Fossil' – taken to be the young De la Beche: 'He guides the helm, whilst by his side a damsel young and passing fair reclines ...' runs 'The Lymiad':[4]

5.1 Henry De la Beche at the age of twenty-three in 1819.

He is a most accomplished youth.
That is, if madam Fame speak truth;
and more than this I cannot tell;
But some who know Sir Fopling well,
inform me he's an F.G.S. [Fellow of the Geological Society].

But this lazy life of privilege wasn't to last. In the early 1830s new legislation banning slavery across the British Empire had begun to hit the profitability of the sugar plantations, and De la Beche was running out of money.[5] Like Murchison, he turned to geology for a new career and rapidly became a respected collector and illustrator of fossils, often with the help of Mary Anning.

As Sedgwick and Murchison began their work in Wales, De la Beche was starting a new job working for the Board of Ordnance, the precursor of the Ordnance Survey.[6] His task was to undertake the first detailed survey of the geology of Devon, Cornwall and west Somerset, using his military training in landscape drawing, topography and surveying to shade in maps of the region using distinctive colours for each rock type. In return he received a one-off fee of £300.

Over the following years De la Beche methodically worked his way north from the limestone cliffs of the South Devon coast. Anecdotal evidence suggested that the region was broadly similar to Wales, with large areas of slate and Greywacke broken up by outcrops of granite and, as he moved north, De la Beche kept an eye on reports of Sedgwick and Murchison's work in Wales, hoping to learn something he could use. By 1834 he had reached the North Devon fishing port of Bideford, where the River Torridge flows between low wooded banks into the open water of the Bristol Channel. Here he found a small mining industry working seams of a shiny black coal known as Culm, an impure form of carbon-rich anthracite.

De la Beche traced the seams inland, recording in his field notes, 'old [coal] pits on the outcrops of the anthracite are very common so the beds may be easily traced across the country'. The seams were mostly narrow and faulted, and many of the pits had been abandoned almost as soon as they had been opened.[7] But what surprised him was not the frequency of the abandoned pits, but the fact that the coal seams appeared to be sitting in the heart of the Greywacke and, equally surprisingly, that this Greywacke contained fossil fragments of marsh-living ferns and horsetails normally associated with much younger

Carboniferous rock. According to Murchison's work in Wales, neither of these observations should have been possible – all Murchison's fossil evidence suggested the Greywacke had been deposited at a time when life on Earth was entirely marine. Indeed, several years earlier Murchison had expressly told a meeting of the Geological Society that there were no land-based plants in the Greywacke, and that it was therefore a waste of time to prospect for coal in Greywacke regions.

The findings were so unexpected, and the fossil fragments so poor, that De la Beche wondered if he had made a mistake. He shipped the fragments off to John Lindley, professor of botany at London University and co-author of the recent authoritative *Encyclopedia of Plants*.[8] Lindley's response was unequivocal: the fossils were identical to well-known plant remains found in Britain's Carboniferous coal fields. Murchison must be wrong about the Welsh Greywacke.

In December 1834 De la Beche's discovery was formally announced to a meeting of intrigued members of the Geological Society. He couldn't afford the cost of the stagecoach to London, so he outlined his findings in a letter to his friend, the mapmaker George Greenough. In this letter De la Beche confirmed that 'the anthracite and fossil plants ... definitely came from an integral part of the Greywacke'.[9]

Murchison wasn't going to let this go without a fight. When he had outlined the details of the Silurian System earlier that summer he had quietly overlooked the existence of the Devon Culm.[10] De la Beche's very public presentation of both the coal and the fossils now rendered this strategy impossible. Murchison didn't doubt that Lindley was right and that the fossil fragments came from Carboniferous plants; what he simply couldn't believe was that they had come from within the Greywacke – De la Beche must have made a mistake. 'Your Bideford paper gave rise yesterday evening to a very animated discussion,' Greenough reported back to De la Beche the following day:

> Murchison led the attack and expressed his astonishment that so experienced a Geologist [as you] should have fallen into so great a mistake as to fancy that the specimens on the table had anything to do with Transition rocks. As no one offered to speak on the other side I ventured to affirm that you had not committed the mistake attributed to you, adding that Murchison had confessedly never seen the country, & that you had examined it attentively, that it was not our practice to give credit to the results of abstract reasoning when opposed to actual observation.[11]

The meeting, according to one account, became 'so animated that the guests may well have felt that the Society's celebrated custom of unbridled argument had got out of hand'.[12] Several days later one of the Society's secretaries, Edward Turner, wrote to De la Beche, 'your discovery of coal and its fossils in the Greywacke was last Wednesday turned topsy-turvy without a scruple ... Murchison in fact said that he had never seen the Country, but would nevertheless take upon himself to affirm, so identical were the fossils with those of the ordinary coalfields, that your coal also belonged to the Carboniferous beds.'[13]

To Murchison, the finding was not merely an academic issue: there were economic implications. He had begun to take pride in the fact that he could warn landowners with estates on the Greywacke – many of them his friends – that they would be wasting their time prospecting for coal. Yet here in the west of England coal and Greywacke seemed to go hand-in-hand, vastly extending the areas of Britain where it might be worth prospecting and the number of landowners who might be sitting on a fortune. Days later, Murchison wrote to De la Beche emphasising that all his years of work in the Greywacke had:

> not produced a fragment of any vegetable approaching to those of the Carboniferous era ... The case you have observed has presented [the rocks] in a deceitful form & may have misled even so good a Geologist as yourself. If your case be substantiated, away go all rules of mining based upon geological and zoological indications. Forthwith may the Welshmen, whom I have checked in their 'decies repetita' coal boring mania, again resume their pickaxes & pierce into the depths of Snowdon in search of the Biddeford Coal.

This put De la Beche in a difficult position. Murchison had become an influential figure in geological circles, and his suggestion that De la Beche might have missed the fact that the Culm sat in a band of Carboniferous rock sitting on top of the Greywacke seriously threatened his reputation as a competent geologist. Yet now, more than ever, he needed to be taken seriously: the Jamaican sugar estates were bringing in less and less income, and the fee from the Board of Ordnance wasn't covering his cost of living. At the age of thirty-eight, De la Beche was running out of money. He would either have to retire to Jamaica to eke out his collapsing income or find a more permanent source of cash in Britain, preferably as a permanent surveyor with the Board of Ordnance. Murchison's broadside called into question his abilities at a time when he needed all the credibility he could muster.

And so he made several anxious return trips to Bideford to check his findings. The established stratigraphic table suggested that between the Greywacke and the coal-bearing rocks there should be that familiar bed of Old Red Sandstone, plus beds of Carboniferous Limestone and a seam of so-called Millstone Grit or, if they were missing, at least signs of unconformity. But as De la Beche examined the rocks around the Culm beds there was no sign of any of them. The strata of Culm containing the plant fossils appeared to sit firmly and squarely in an unbroken, conformable sequence of Greywacke rock. He had not made a mistake.

This was a blow to more than simply the integrity of Murchison's Silurian System. If the rocks in Devon really did have a different fossil fingerprint from those in Wales, then they threatened to undermine Murchison's entire attempt to construct a fossil-based stratigraphy. Some weeks later Murchison wrote yet again to De la Beche, expressing his concern. 'It is a sequitur of the establishment of your inference,' he pointed out, 'that strata can no longer be identified by their fossil remains.' And Sedgwick, always less committed to the role of fossils in stratigraphy, wrote to De la Beche, with perhaps a hint of enjoyment, 'The fossilologists are going mad ... [Murchison] has, as you know, examined an astonishing fine series of upper fossiliferous Grauwacke and finding no coal plants in it, resolved in his own mind that the said plants could not be found any where else.'[14]

In July 1836 Murchison decided to take a look for himself. He persuaded Sedgwick to join him. After several days on the road the two men arrived in the faded North Devon fishing port of Minehead.[15] They paused briefly to explore the ridge of hills rising above the town, climbing steep zigzag paths through dense woods of beech and oak, and declaring them all 'old Greywacke' or Cambrian, before pushing west across the open moors of the North Devon coast. The narrow road is cut into the hillside as the land falls steeply into the Bristol Channel and traces the line of some of the most spectacular cliffs in the country. On the horizon the hills of south Wales appear across the water.

The two men carried with them the local sheet of George Greenough's 1820 map as well as De la Beche's more recently published large-scale local maps. Murchison was openly dismissive of them: 'this portion of De la Beche's map wretchedly worked out,' he grumbled, 'half the limestone [is] omitted'.[16] But De la Beche's maps showed in detail the location of quarries, road cuttings and coastal cliffs and enabled the two men to cover the ground quickly. The plan was to confirm rapidly the identity of the rocks, usually on the basis of their physical appearance, collect whatever fossils they could for later study and move on as fast as possible. It was a rough-and-ready approach, but neither man was planning a detailed survey of the region's geology; their main concern was to align it with their work in Wales.

Murchison was anxious to prove that De la Beche had made a fundamental error and then return quickly to London to concentrate on developing his Silurian empire. Sedgwick was less concerned about Carboniferous fossils in the Greywacke, but was under growing pressure to back his stratigraphic classification of the Cambrian strata with at least the outline of a fossil fingerprint. Given the almost complete absence of Cambrian fossils in Wales, he fervently hoped that the apparently equally ancient rocks in Devon and Cornwall would yield a better harvest. After their successful mapping of the Welsh rocks, neither man was expecting it to be a major challenge.

They stopped briefly in the small North Devon ports of Lynmouth and Ilfracombe, where large new hotels offered the latest conveniences, while the terraces of handsome guesthouses vied to attract the more adventurous Victorian holidaymaker. The Napoleonic Wars had driven wealthy British families, who traditionally holidayed on the Continent, to seek out new British resorts; and the small Devonshire fishing ports were regarded as an 'agreeable summer residence' and 'bathing-place'.

Beyond Ilfracombe the land rose and fell. In sandy bays and inlets, where the sea broke against the gritty limestone cliffs, the geologists found buried in the rock an 'abundance of Encrinite heads', better known as crinoids or sea lilies. Murchison had found similar fossils in the Caradoc Sandstone of the Shropshire Hills, which suggested these rocks might belong to what he was calling his Lower Silurian. But their coarse, gravelly texture and appearance looked more similar to the Cambrian strata in Wales, and he reluctantly concluded they were more likely to be 'very ancient Cambrians'.[17]

They continued round a coast of wide sandy beaches and low dunes to the Taw Estuary where, on the northern outskirts of Barnstaple, they came across an exposure of black shelly limestone that looked remarkably similar to Murchison's Llandeilo Flags, suggesting that they were passing from older Cambrian rocks along the northern coast into younger Silurian rocks further south around Barnstaple.[18] So far, so good: the rocks of Devon seemed broadly to fit into their Cambrian and Silurian classification.

On the southern outskirts of Barnstaple, however, the track turned black and dusty and they came face to face with De la Beche's troublesome Bideford Culm. Fingers of smooth, black, shiny rock ran out into the sea. Murchison made a quick inspection and concluded that it was 'wholly unlike anything in the Greywacke Series' and bore all the signs of being Carboniferous. In stratigraphic terms, it meant that somewhere during the journey from the north to the south of Barnstaple – a distance of a little over 5 miles – they had passed from Lower Silurian Llandeilo Flags into much younger Carboniferous coal measures.

Yet they had seen no sign of the strata that normally represent the tens of millions of years that separate these two periods, or of an unconformity marking where they should have been. As they continued south out of Barnstaple, along gravel tracks 'so narrow that a mouse cannot be passed by any carriage',[19] they comforted themselves with the thought that the unconformity probably lay somewhere under the water and mud of the nearby Taw Estuary.

To Murchison, a man inclined to sweeping conclusions, the picture now seemed clear. The rocks to the north were older Greywacke, mostly Cambrian, and appeared to dip to the south. According to Greenough's map, the rocks of southern Devon were also Greywacke. Putting the two together suggested that the Culm must be lying in some kind of Carboniferous trough, or syncline, nestling in a bed of older Greywacke, with the northern boundary of the Carboniferous trough lost in an unconformity somewhere under the Taw Estuary. Murchison sketched a rapid cross-section through the region, highlighting the unconformity and the trough of Carboniferous rock. 'The truth I can only surmise,' he would later write, was that De la Beche, 'had never really looked carefully at the consecutive sections [or strata] in nature.'[20] All that remained to be done was to track down the southern boundary of the trough in order to confirm his argument.

Over the next few days the two men followed the North Devon coast into Cornwall as far as the cliffs at Tintagel, a wild, empty landscape. Here, however, the rocks were unexpectedly difficult to read. Notwithstanding Murchison's earlier assertion that the Culm was 'wholly unlike' the Greywacke, they found it extremely difficult to tell whether they had passed out of the 'Carboniferous' trough and into the southern Greywacke rocks. 'From the Carbonaceous appearance of the shales & the little difference of lithological structure between the Sandstones on this coast of Cornwall & those of Bideford ... we still imagined we were in the deposits of the Carboniferous age,' wrote Murchison.[21]

To confound things further, as they explored the coastal cliffs at low tide they found the rocks contorted into extraordinary troughs and saddles, the strata often plunging vertically into the ground. At Bude they were pinched up into a series of small arches or anticlines; just south, at Millook Bay, they formed a succession of chevron-like zigzags. It made it almost impossible to determine the chronological order of the strata.

Finally in some confusion – uncertain whether they were still in the Culm trough or had crossed into the southern Greywacke – the two men turned inland through a landscape of big skies and small oak woods until, near a small cluster of stone houses known as South Petherwin, they found a band of fossil-rich limestone. It ran through a quarry that looked remarkably similar to the limestone cliffs along the northern coast of Devon which they had assigned to the Cambrian, suggesting that they had finally passed the southern boundary of the Culm trough and were back in the older Greywacke. Yet once again the boundary itself had completely escaped them.[22]

They continued south along narrow lanes with high blackthorn hedges, exploring what they assumed was the southern Greywacke as it stretched down to the cliffs of the South Devon coast. Greenough's map simply marked the entire area as Greywacke, but since there were no signs of anything Murchison recognised as Silurian rock, they calculated they were now deep in Sedgwick's Cambrian. For Sedgwick, this was an important discovery. The cliffs and rocky outcrops around Plymouth and the Torbay area were packed with fossils: if they really were Cambrian, they represented his best opportunity yet for identifying an authentic Cambrian fossil fingerprint. But the designation was uncertain. De la Beche had marked the cliffs as Carboniferous rather than Greywacke, and the fossil record included species commonly found in Carboniferous rocks.[23]

As the two men scoured the cliffs for fossils, rain swept in from the English Channel and for several days they were forced to take refuge while Sedgwick worried about his health. 'We are in shelter, with a tea urn hissing on the table, and some broiled fish within nostril scent,' he reported from Plymouth to a friend.[24] He also recorded in his field notes a recipe for a drink to drive out the cold: a cocktail of opium, cardamom, peppermint and ginger.[25] It didn't seem to work, and he later reported suffering from 'severe attacks [of flu] brought on by fatigue and bad weather'.[26]

At last, after a week cooped up together, the skies finally cleared. Sedgwick rose from his sickbed and the two men began the return journey to the northern coast. They skirted the eastern edge of the granite mass of Dartmoor and finally, near the large village of Chudleigh on the eastern flanks of the moors, found a quarry where, tucked into the bands of newly exposed mud and limestone, they could see the merest suggestion of an unconformity that might be regarded as marking the elusive southern boundary.[27] With nothing else to go on, it was enough for Murchison to convince himself that his earlier thesis of a trough of Carboniferous Culm sitting unconformably on a bed of Greywacke was an accurate representation of the region's geology.

Several days later, after little more than four weeks of fieldwork, the whistle-stop tour of the region was complete and they returned to Bristol to attend the sixth annual meeting of the British Association for the Advancement of Science. Here Murchison presented a rough-and-ready outline of their findings.

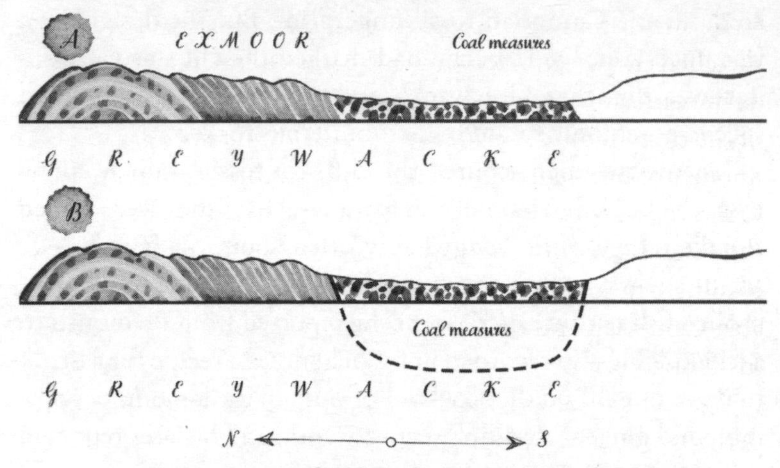

5.2 A comparison of De la Beche's cross-section through Devon (A) with Sedgwick and Murchison's (B).

'The chief business of Sedgwick and self,' he later noted, 'was to establish the point regarding the great change we proposed to make in the structure of Devonshire.'[28] Word had already leaked out that he would be challenging De la Beche's view of the Culm as an inseparable part of the Greywacke and there was considerable excitement. 'Sedgwick and Murchison project a fierce encounter,' noted De la Beche's friend George Greenough on the eve of the meeting.

Murchison's presentation consisted of a series of sweeping conclusions that differed fundamentally from De la Beche's work.[29] The latter's map of the region suggested that the rocks of Devon formed an unbroken sequence of south-sloping strata that passed from Greywacke in the north to Carboniferous in the south, with the Culm forming an integral and unbroken part of the succession. Murchison's map, by contrast, argued that the Culm represented a trough of Carboniferous rock sitting in a bed of Greywacke to its north and south.

It was a hurried piece of work, and Murchison's rapidly sketched cross-section obscured the lack of any clear evidence of an unconformity on either the northern or southern boundaries of the trough. But for both Sedgwick and Murchison it solved a number of problems. For Murchison it meant the troublesome Culm rocks, with their Carboniferous fossils, were well clear of the Greywacke. In Sedgwick's case it suggested that the richly fossiliferous limestone cliffs along the southern coast were probably Cambrian and therefore a potential source of a Cambrian fossil fingerprint.

At the end of the meeting De la Beche complained, with perhaps only the semblance of good humour, 'I was taken most deucedly in the flank … My ammunition being in my magazine and my guns dismantled, expecting nothing but peace, I made my retreat in the best manner I could.'[30] But the retreat left him in a difficult position. The previous year he had finally secured a permanent position as the first director of a newly established geological branch of the Ordnance Survey – the Geological Survey of Great Britain. It paid a relatively handsome annual salary and provided him with a number of assistants. Sedgwick and Murchison's paper suggested he had made a fundamental mistake that might jeopardise this new-found security.

He needn't have worried. Over the following months further fieldwork only confused the picture and Murchison's neat outline of the region's rocks gradually fell apart. Local geologists, spurred on by the high-profile interest in their region, turned up examples of Culm-type fossils outside the area mapped by Murchison, suggesting that the borders of the Culm trough were not as clear as he had depicted them to be.[31] Doubts also began to arise about the designation as Cambrian of the South Devon limestone cliffs around Plymouth. Robert Austin, a Devonshire landowner who had studied geology under William Buckland at

Oxford and assembled one of the best collections of local fossils in the region, announced that he had discovered more than thirty fossil species in the cliffs around Tor Bay and Newton Abbot that looked 'strangely analogous' to well-known Carboniferous species.[32] It seemed to confirm De la Beche's designation of the rocks as Carboniferous and to further undermine Sedgwick and Murchison's map of the region, which depicted the whole of South Devon as Cambrian. It was particularly unwelcome news for Sedgwick, who still clung to the hope that the area might furnish him with a Cambrian fossil record.

But most puzzling of all, new questions arose about the nature of the Culm itself. De la Beche's original specimens had been poor and badly damaged. New and better examples had now been unearthed, and John Lindley, who had sparked the whole controversy two years earlier by identifying them as Carboniferous, now reversed his opinion. He re-examined the old specimens, compared them with new ones and declared that most had absolutely nothing in common with any Carboniferous fossils he had ever seen before. What could that mean? If the Culm was Carboniferous, yet contained its own unique fossil fingerprint, then once again the whole attempt to establish a universal fossil-based stratigraphy was thrown into question. But if they weren't Carboniferous, what were they? They didn't seem to be Cambrian or Silurian. In the words of Charles Lyell, currently serving his term as president of the Geological Society, the 'Culm question' was turning into 'one of the most important in a theoretical point of view, ever discussed at the Geological Society'.[33]

In the winter of 1837 the British Association, in an attempt to finally thrash out an answer, agreed that Sedgwick and Murchison should present their work to a meeting of experts in London. Murchison worked hard on his notes. Sedgwick promised to supply his thoughts. Back in Norwich, however, swamped by cathedral duties and 'feeling depressed by the cold formality'[34] of the services, he failed to put pen to paper. Days before the meeting Murchison was still waiting for his contribution. 'I was in great hopes to have your despatches before now,' he wrote in a quick note to Sedgwick. 'But I wait patiently like a lamb for the sacrifice, and sacrificed I most assuredly shall be without your aid. However, I will drink the best part of a bottle of sherry to screw me up to face ... [the forces] which are to be brought against us.'[35]

The notes never came and the meeting was abandoned. 'The part of Hamlet omitted, the play was not performed,' Murchison wrote to Sedgwick two days later. 'I am mortified that the memoir did not come ... did you really imagine that I was to dramatise the whole thing without a sermon before me?'[36]

A year later, in May 1838, Sedgwick finally produced a paper on his Cambrian rocks. Designed to complement Murchison's earlier paper on the Silurian, it was introduced, like most important geological announcements, at a meeting of the Geological Society in London, but in other respects it was not what his fellow geologists had hoped for.

'A Synopsis of the English Series of Stratified Rocks Inferior to the Old Red Sandstone' was a mere eleven pages long and dealt only briefly with Wales, repeating Sedgwick's earlier division of the Cambrian into four periods: the Upper Cambrian slates of the Berwyn Hills and southern Wales, the Bala Limestone, the Middle Cambrian slates of Snowdonia and the Lower Cambrian schists of Anglesey and the Llyn Peninsula. But it ducked any attempt to assign a definitive fossil fingerprint to any of them – largely because Sedgwick still hadn't found one.

To most of his fellow geologists, now largely committed to

Murchison's concept of a simple fossil-based stratigraphy, it was a serious disappointment. It contained none of the bold clarity of Murchison's Silurian System and, without a series of clear and unique fossil fingerprints, it failed, in most people's eyes, to identify the Cambrian as a convincing universal stratigraphic category that could be readily applied across the world.

Sedgwick sought to defend his position. 'When a system has been well made out, and its groups of fossils determined, we may then make use of comparative groups of fossils freely, and with very small risk of mistakes,' he wrote. 'But to begin with fossils, before the physical groups are determined, and through them to establish the nomenclature of a system, would be to invert the whole logic of geology and could produce nothing but confusion.' In Wales, he argued, it was impossible to identify the individual strata on the basis of the very few fossils he had found. In Devon he was more or less resigned to the loss of the fossil-rich limestone cliffs of the south coast, but still clung to the hope that the limestone rocks near the small hamlet of South Petherwin, which he had explored with Murchison in 1836, might yet provide him with a unique and worthwhile fossil record. It was probably the best he could do, but few people were convinced.

UPPER CAMBRIAN	Higher beds of Berwyns/south Wales slates
	Bala Limestone
MIDDLE CAMBRIAN	Higher mountains of Caernarvonshire and Merionethshire, including Snowdonia
LOWER CAMBRIAN	The mica schists of Anglesey and south Caernarvonshire

5.3 Sedgwick's Cambrian stratigraphy.

Murchison hammered home his advantage. Six months later, in January 1839, he published his 'big book', *The Silurian System*, in two large quarto volumes, costing eight guineas each (about £500 today); six double-column pages listed the names of subscribers who had contributed to the cost of its publication. It was an extraordinarily thorough piece of work, running to 800 pages, with dozens of maps, cross-sections and fossil plates. A fulsome dedication noted, 'My dear Sedgwick ... having explored with you many a tract, both at home and abroad, I beg you to accept this offering as a memorial of friendship, and of the high sense I entertain of the value of your labours.'

Volume 1 ran through the Secondary rocks of Britain, chapter by chapter, from the 'Oolitic System' (the limestone of the Cotswolds) through the Carboniferous to the Old Red Sandstone. It didn't reach the Silurian System until Chapter 16, which began, 'we have at length reached those older deposits which having not been separated into formations by previous writers I am compelled to describe under new terms'. It then reiterated what Murchison had already outlined, dividing the Silurian into an Upper and Lower system with clear breaks, or unconformities, between the Silurian and the geological systems above and below it.

Volume 2 continued in the same exhaustive manner, laying out the Silurian fossil record and reiterating Murchison's belief in the central importance of a fossil-based stratigraphy. To some readers, the detail was a little excessive. 'Did your eyes ever light upon such a mortal mass of heavy writing and dry profitless mineralogical detail as that contained in Part 1 of *The Silurian System*,' wrote the Devonshire geologist Robert Austin to his friend De la Beche.

Despite its off-putting bulk, the book was an immediate success. Copies of *The Silurian System* were sold across Europe and North America. From France to southern Africa, geologists began identifying Silurian rocks using Murchison's list of fossil fingerprints. John Bull seems 'to lord it over Europe in geology,' George Greenough observed, and there was a minor backlash among some geologists worried about the universal applicability of a small sequence of rocks along the Welsh–English border. 'I can foresee the fate of geology for the next eight years – half the globe will become Silurian,' Robert Austin commented tartly.[37]

As it happened, the most thorough application of Murchison's thinking was in Germany, where an eccentric German geologist called Christian Leopold von Buch began to identify rocks across a wide area of Germany and Eastern Europe as Silurian. Von Buch was a wealthy Berlin bachelor with a reputation for taking long, meandering walks through the German countryside, dressed in an overcoat fitted with outsized pockets, which he stuffed with notebooks, maps and geological implements. 'He had rambled all over Europe everywhere making the acquaintance of those who followed the same pursuits,' wrote one observer. 'Probably no geologist of his time had so wide a range of knowledge and acquirement. He was great and original in physical geography, in dynamical and stratigraphical geology [and] in palaeontology.'[38]

Von Buch had heard about the concept of the Cambrian and Silurian systems in the wake of the now-famous paper that introduced the terms at the 1835 meeting in Dublin. Since then he had raced ahead, using Murchison's list of fossils, categorising rocks as Silurian more boldly than even Murchison had so far dared. Within a year or so von Buch had identified as Silurian an enormous area stretching from the rocky crags of the Harz Mountains in northern Germany through the plains of Poland to the then-border with Russia, with patches around the Czech capital of Prague and the Russian capital of St Petersburg as probably older Cambrian rock. He even claimed to have identified a single fossil species, a genus of brachiopod called Spirifer,[39] which was unique to Silurian rocks and clearly distinguished them from Cambrian strata – a marker fossil that neither Sedgwick or Murchison had been aware of. Almost inevitably, the Silurian System began to overshadow the Cambrian and highlight Sedgwick's failure to produce such an equally straightforward guide to the rocks; as one historian has written, 'Fossiliferous areas previously compared to the Cambrian were now ceded to Siluria.'[40]

Yet the success of Murchison's book hid a number of shortcomings. It made no reference to any uncertainty surrounding the Cambrian/Silurian boundary, or to the arrangement of convenience that had given rise to it, and ignored any of the new fossil evidence coming from Devon and Cornwall, which appeared to fudge the boundary between the Greywacke and the Carboniferous systems above it. Indeed, it was remarkably vague about the upper boundary of Siluria. In one section, Murchison reaffirmed his belief that there had been an abrupt environmental change between the Silurian and Carboniferous systems, leading to a clear break in the fossil record. Elsewhere, however, he made much of his early discovery of 'passage beds' along the Wye Valley, where he had found 'an unbroken transition from the Mountain [Carboniferous] Limestone downward through the Old Red Sandstone into the fossiliferous greywacke'.[41] In other words, a gradual rather than abrupt boundary.

Above all, his account glossed over the problems raised by rocks like the Devon Culm. These rocks, with their idiosyncratic Carboniferous fossil record, suggested that Murchison's approach to the fossil record was just too simplistic to reflect accurately the complexity of the evolution of the Earth.

One of his sharpest critics was John Phillips, the thirty-nine-year-old nephew of the celebrated mapmaker William Smith. In his recently published popular *Guide to Geology*, Phillips, now professor of geology at King's College London, argued it was inconceivable that a single geological period covering millions of years, and encompassing the entire globe, would contain only one physical and climatic environment and would therefore leave one fossil fingerprint. Likewise, there was no reason why two different geological periods might not contain similar environments and therefore a common fossil record. As one commentator put it: 'flora of so-called Coal Measure species might have flourished in specific ecological conditions during the Greywacke period long before it was able to spread widely and become the "characteristic" flora of the Coal Measures period.' A comprehensive stratigraphy, Phillips argued, had to be much more subtle and sophisticated. It should include important indicators such as the structure of the rocks – their relationship to each other – and their composition and characteristics.

Murchison had a short, though somewhat mystifying, answer to this objection. He suggested that the early Earth had been hotter, and had therefore produced a more uniform climate across the globe. His friend Charles Lyell also developed an elaborate theory to explain how the Earth's ever-changing flora and fauna could be linked to a history of changing physical environments, and concluded that two geological formations, deposited at different times, would be most unlikely to share common species.[42] Neither bore serious investigation, but with most geologists increasingly invested in the idea of Murchison's fossil-based stratigraphy, they went largely unchallenged.

And so Sedgwick faced a choice. If he was to continue to argue that Cambria represented a unique, universal, stratigraphic period in the Earth's history, he could do one of two things: either pull together a fossil record that clearly justified the claim, or follow Phillips's example and produce a more sophisticated stratigraphy that developed his well-known insistence on the importance of the structure of the rocks.

Unfortunately he did neither. Friends urged him to get on with it. 'Do not refuse or get into a rage because I say you can do this,' Murchison wrote to him. 'You would thoroughly write it in fifteen days. The whole of the matter is completely in your head.' The newly serving president of the Geological Society, the Cambridge philosopher and scientist William Whewell, expressed the hope that Professor Sedgwick would not 'withhold from the world much longer the views of his sagacious and philosophical mind'.[43] But Sedgwick, never the most enthusiastic collaborator, was struggling to translate his discursive abilities onto the written page. 'Some of my friends expect more from me than I can do. Because on some occasions I speak fluently, it by no means follows that I write readily. Experience tells me the very contrary.'[44]

He made various excuses: 'My lectures begin the day after tomorrow and I have three papers on the stocks. My hands were never so full. I wish the days were forty-eight hours long or that I could find some patient mode of getting through the term without eating or sleeping.'[45] He was also obsessed with the effects of writing on his health. 'I write with pain to myself in every sense of the word, for a very few hours writing will bring on a fit of indigestion and a swimming in the head,' he noted to a friend.[46] 'Could I command my stomach, I could clear off my work easily enough; but the moment I try to employ my pen, all the inner man gets out of order, and then, as for the brain, my head might just as well be filled with brick-dust.'[47]

'A connected argument of substantial length seemed beyond his powers,' writes the historian of geology James Secord. 'It was frequently difficult,' wrote another observer, 'to distinguish between actual bodily ailments and bouts of hypochondria, brought on by long periods of depression.'[48] To a driven man like Murchison, ambitious and determined to push ahead, it must frequently have felt like a pointless collaboration. He had sent early drafts of *The Silurian System* to Sedgwick for comment, but received little useful feedback except a stinging rejection of his claim to have 'discovered' the Silurian in the Wye Valley. 'I must honestly say I don't like the tone of your introduction,' Sedgwick told Murchison in a letter. 'You assume a prescience I don't believe real ... and you set down as little better than a pack of asses everyone who had preceded you.'[49]

Sedgwick later proved equally unwilling to review or even read *The Silurian System* once it was published, declining a request to review it for *The Times* on the grounds that he had no time.[50] It's hard to escape the sense that he was beginning to resent the success of his collaborator.

In February 1839, a month after the publication of *The Silurian System*, De la Beche published his own account of the rocks of the West Country of Britain: *A Report on the Geology of the West Country of Cornwall, Devon and West Somerset*.[51] This publication boasted none of the cachet of *The Silurian System*. 'Whereas Murchison's elegant volume was designed to grace the libraries of the nobility and gentry of "Siluria" and charm its subscribers with geologically orientated vignettes of their familiar local scenery,' writes Martin Rudwick in his definitive account of the Devonian controversy, 'De la Beche's book was altogether more forbidding.'[52] In 700 densely packed pages, it summarised at huge length his existing position that there was no evidence of a Carboniferous trough sitting unconformably in the Greywacke, and continued to insist that the Culm was probably the equivalent of Murchison's Upper Greywacke or Silurian rocks.

As Murchison skimmed through an early copy, his impatience mounted. 'I really think there is nothing in the history of science,' he thundered in a letter to Sedgwick, 'so bare-faced & unprincipled & ungentlemanly like as this Spoliation.' He fulminated at the idea that public money, in the shape of the Geological Survey, should have been spent helping to produce it.[53] Yet despite his bluster, it was becoming clear there was simply no evidence for Murchison's supposed unconformity across North Devon, which was meant to separate the Culm from the Greywacke. He was also troubled by the recently discovered fossil anomalies in the Culm, which seemed to suggest it was not typical Carboniferous rock.

It was now that Murchison, always a big-picture man, arrived at one of his boldest conjectures. For some time he had been eyeing the Brecon Beacons and the colourful Old Red Sandstone hills to the north of the Bristol Channel and wondering whether they stretched further south. Could it be that they ran under the Channel, to re-emerge as the North Devon cliffs: rocks that, until now, he had assumed were Cambrian? That would mean the 'Carboniferous' Culm beds around Barnstaple, rather than sitting unconformably on the Greywacke, sat conformably on top of the Old Red Sandstone. In one sweep it removed the need to posit an unconformity running through the region. Moreover, if the rocks of the area were indeed a mix of Old Red Sandstone and Carboniferous Culm, then the anomalous fossils turned up by local geologists were probably transitional species caught in the change from the Old Red Sandstone to the Carboniferous period.

Murchison put his new thoughts on paper. Hitting his stride, he floated an even bigger idea. Until now the Old Red Sandstone had been seen as a distinct stratum of rock lying at the bottom of the Carboniferous beds, perhaps even part of them. Murchison now suggested that, far from being a sub-section of one of the better-recognised geological periods, it was a geological period in its own right: a hitherto unrecognised phase in the evolution of life, in the course of which the almost exclusively marine flora of the Silurian Period had slowly given way to the land-based ferns and tropical plants of the Carboniferous. He called this transitional period the Devonian.

Geologists around the world were intrigued. Although the idea was based, in the words of one commentator, 'rather on what seemed probable than what had been proved to be the case', it was an ingenious solution to an intractable problem. The early rocks of the Earth could now be divided into four great systems, rather than three: the Carboniferous, already internationally recognised; the Silurian, which since the publication of *The Silurian System* was well on its way to being internationally recognised as a universal system; the Devonian; and the Cambrian.[54]

BASIC CATEGORY	GEOLOGICAL PERIOD
SECONDARY ROCKS	Cretaceous
	Jurassic
	New Red Sandstone/Triassic
	Carboniferous
GREYWACKE	DEVONIAN
	Silurian
	Cambrian
PRIMARY ROCKS	

5.4 The newly identified Devonian Period came to constitute a third distinct period in the early history of the Earth.

In the winter of 1840 Sedgwick and Murchison's new geological period won the endorsement of William Buckland, about to serve his second term as president of the Geological Society. In his anniversary address to members he confirmed his support for a tripartite division of the Greywacke into Devonian, Silurian and Cambrian systems and praised it as 'undoubtedly the greatest change which had ever been attempted at one time in the classification of British rocks'.[55] International support from geologists in France, Germany, Belgium and the US was not far behind.

Yet the immediate success of the Devonian solution obscured a number of concessions that Murchison had made to his critics. The Old Red outcrops in the Brecon Beacons and in the sea cliffs along the coast of north-east Scotland were composed of a brightly coloured sandstone with a very distinct fossil fingerprint: a series of fish fossils, and in particular the marker fossil Holoptychius.[56] The rocks in Devon and Cornwall were composed of a much paler limestone and contained a completely different fossil record.[57] In proposing the existence of a Devonian period, Murchison was therefore violating one of his cardinal convictions: that fossils were the key to identifying rocks. The new formulation forced him to admit that opponents such as John Phillips might be right – a single geological period might be characterised by local variations in the fossil record.

Equally importantly, Murchison's proposal of the Devonian undermined an argument, made only months earlier, that geological periods were usually marked out by unconformities: sharp breaks in the fossil record. The Devonian, with its changing marine fauna and gradual appearance of land-based plant fossils, represented a gradual faunal transition from Silurian to Devonian and from Devonian to Carboniferous. In other words, it suggested that – unlike the vivid banding depicted in stratigraphic tables – in the field the boundaries between different geological periods could be subtle, with the fossil record only gradually changing from one to another.

Finally Murchison faced a basic structural problem. The rocks of Devon and Cornwall were so twisted and disturbed there wasn't a single outcrop that clearly showed the Devonian sitting in a sequence between the Silurian and Carboniferous. Just as some years earlier the absence of the Old Red in north Wales had left the Greywacke of the region floating in a temporal vacuum, so the absence of a clear Silurian–Devonian–Carboniferous sequence meant that Murchison could only speculate that it existed.

Then in early 1840 he received a letter from his friend in Berlin, Christian Leopold von Buch, who had done so much to spread the Silurian System across continental Europe. Von Buch told Murchison that he had started deploying the Devonian as a stratigraphic category and had found large outcrops of the rock across areas of western Russia near the Polish border. Furthermore, these rocks lay in orderly, almost horizontal strata, which plainly demonstrated that the Devonian rocks sat neatly sandwiched between the Silurian and Carboniferous strata, exactly as Murchison had predicted. Furthermore, von Buch claimed that he had found specimens of the fish fossil Holoptychius, characteristic of the Old Red Sandstone, in close association with some of marine shells and corals that were common in Devon and Cornwall: the first time the two sets of fossils had been found together in one piece of rock. The letter concluded by suggesting that Murchison should make a field trip to the region to confirm the observations.

As Sedgwick continued to struggle to clearly define his Cambrian System and its relation to rocks elsewhere in the world, Murchison was about to undertake a remarkable series of journeys that would turn him into perhaps the world's most influential geologist.

Chapter Six

STUMBLING ON THE PERMIAN IN RUSSIA, 1840–42

In May 1840 Roderick Murchison set out for western Russia. He took a paddle-steamer to the German city-state of Hamburg across a heaving North Sea, with black smoke billowing from the funnel and the boat heeling to one side, 'one paddle for the most part only at work'.[1] From there he continued by carriage to the Prussian capital of Berlin. His companion on this journey was a young French palaeontologist called Édouard de Verneuil, a wealthy Parisian lawyer who had given up the Bar to follow the new science of the rocks and had come to regard Murchison as a mentor. Murchison, in turn, held the Frenchman in high regard, at least partly because of his skill in identifying early Palaeozoic fossils. Murchison and de Verneuil would become inseparable over the next few years.

In Berlin the two men were joined by a Russian-German geologist and naturalist called Alexander von Keyserling, a scion of one of Germany's ancient aristocratic families brought up in the borderlands where Prussia met Russia. Von Keyserling, a lifelong friend of the Russian tsar, Nicholas I, had been recommended by Leopold von Buch because of his familiarity with the geology of western Russia.

The three men travelled north to the Baltic coast and then on by steamer to the Russian imperial capital of St Petersburg.

They had been given permission to join a Russian fact-finding tour of the country's natural history and resources: the plan was to travel 600 miles north-east to the Arctic city of Archangel on the shores of the White Sea, then loop back to Moscow and their starting point, via the central Russian city of Nizhny Novgorod in the Volga valley. The objective was to investigate the extent of the Silurian rocks and the nature of the Devonian.

The party spent a merry three weeks assembling a mountain of equipment, including bedding, wine, medicines, cigars (which Murchison viewed as indispensable), a new culinary invention called tinned soup, and clothes for every conceivable occasion, including full evening dress. Murchison had already bought a lightweight two-wheeled carriage in Germany, but it needed reinforcing to cope with Russia's roads; they supplemented it with a low-slung Russian vehicle drawn by up to five horses, known as a tarantass. They obtained maps, met local geologists and mining experts, hired a domestic servant and winkled out the endless permissions required to journey through a heavily autocratic police state. Permits were needed not only to move from one region to another, but also to leave the highway and investigate the immediately surrounding countryside. Murchison, however, was in his element. Unlike many British geologists, he was comfortable in international circles and spoke good French, the language of the Russian court.

Geological knowledge of Russia was patchy at the time. At the invitation of Catherine the Great, a Prussian naturalist called Peter Pallas had travelled through western Russia and parts of Siberia in the late 1700s, producing notes, maps and a book describing the country's natural history, along with first impressions of some of its geology, which were frequently misleading.[2] More recently the Prussian naturalist and explorer

Alexander von Humboldt had made an extensive tour of western Russia and published a series of maps of the country's natural resources.[3] A British diplomat based in St Petersburg called William Strangeways had also mapped some of the country's rocks. Finally Murchison had brought with him crude maps of the Russian Empire produced by the London-based Society for the Diffusion of Useful Knowledge, an organisation dedicated to publishing inexpensive maps and texts for the lay public.[4]

Murchison was enormously attracted by the grandeur of the Russian capital, with its Imperial Palace and wide, majestic streets. The recently refurbished Winter Palace, he wrote to his wife Charlotte, 'is now in full beauty, and the blaze of light, the elegance of the candelabras, and the masses of gold, quite rivet attention. *We* have no notion of lighting, and I now understand the criticism of the foreigners who attended our Coronation' (emphasis in original).[5] But by mid-June the preparations were complete and the party set off.

The road north out of St Petersburg, like most Russian roads, followed a series of wide, low river valleys through hills of pine and birch. Melting snow and recent rain had caused the rivers to burst their banks. In places the heavily laden party was forced to cross large expanses of deep, muddy swamp. The track rapidly deteriorated. Murchison wrote in his journal, 'our postilion chanted away, as he mercilessly flogged his horses – "go along my beauties! You pull the Silver Gentlemen (a gentle hint to us to pay well), they will give you fine pastures."'[6] In the evenings they stopped at impoverished villages where they 'quartered themselves for the night on some priest or peasant, sleeping generally on their own "shake-downs" upon the floor', or occasionally at an inn, where 'Ivans' (Murchison's generic term for lower-class Russians) served them potatoes and grog.[7]

Murchison couldn't resist supercilious digs at the local

women who, he recorded in his diary, in 'the most unaccountable fashion ... tie down their breasts so as to approach very near to their hips, their stays or ties coming up to their armpits'.[8]

It was high summer. By the light of the midnight sun he spent the evenings disconsolately swatting mosquitoes and, after the excitement of St Petersburg, describing in his journal 'a region of small lakes, fine vegetation and undulations rising to heights of 500 to 600 feet'.[9] For several days the group struggled to find a single exposed outcrop of rock that offered any clue to the underlying geology. And then, just south of the wide, reedy expanses of Lake Lagoda, they struck lucky.

The River Volkhov had cut its way down through a band of limestone, creating a low cliff composed of horizontal bands of red rock. Murchison and de Verneuil leapt on it, knocking off lumps and examining them closely. They could just make out the remains of the brachiopod *Spirifer devoniensis*, a fossil that Murchison had last seen in the cliffs of Devon, and a marker for his Devonian rocks. It suggested that they had arrived in Russia's equivalent.

Even more remarkably, the rocks contained the fossilised scaly outline of the primitive extinct fish Holoptychius, a marker for classic outcrops of Old Red Sandstone in northern Scotland and the Brecon Beacons. It was exactly as von Buch had promised: here were red sandstone rocks that united both the brachiopods of Devon and Cornwall and the fish fossils of the Scottish Old Red. Murchison was ecstatic. Within days of setting out he had achieved one of his primary objectives: to confirm that the sandstones of the west of England and northern Scotland belonged to the same geological period – a crucial validation of his Devonian System. 'A great triumph for myself and Sedgwick,' he wrote joyfully in his journal:

> When we contended that the limestones and sandstones of Devonshire were of the same age as the Old Red Sandstone of Scotland ... We were met with objections – show us a fish of the Old Red Sandstone in Devon or a Devonshire shell in the Old Red of Scotland. Here, in Russia, I had solved the problem; for these shells and these fishes (species for species) were unquestionably united in the very same flag-stones.[10]

Further along the river there was another important find. Drifting downstream in a small boat to inspect the overhanging rocky banks, the party shot a set of rapids where the fast-flowing water had cut its way into the reddish limestone and exposed a series of paler limestone ledges just above the water line. Breaking off samples, Murchison could clearly see that embedded in them were traces of the sea urchin Echinosphaerites, a fossil that he had used as a marker for his Upper Silurian rocks in Wales: proof of the Devonian rock's stratigraphic position immediately above Silurian limestone. The banks of the Volkhov had exposed the elusive stratigraphic boundary between the Silurian and Devonian periods that he had always speculated must exist, but had never been able to find in the contorted rocks of Devon and Cornwall. In a matter of days, Murchison and de Verneuil had cracked the two major issues dogging the Devonian System: the unity of the Old Red Sandstone in Scotland and the rocks of Devon and Cornwall and their stratigraphic position immediately atop the Silurian Period. It was more than he could reasonably have expected.

The expedition rolled north in high spirits. As they pushed further beyond the Arctic Circle the trees became smaller and more stunted. They passed muddy little villages clustered around small wooden churches and whitewashed monasteries. In one of these settlements they stumbled upon a party of political exiles en route to a labour camp in the far north, each man dressed in a coarse grey smock and chained to his neighbour. Murchison, blind to Russian authoritarianism, felt little sympathy for the country's political dissidents. In another settlement, buried deep in the trees, the inhabitants feted the 'official' visitors with music and dancing. It resembled 'a sort of highland fling and shuffling hornpipe ... in a minor squeaking key with nasal chanting'.[11]

'I cannot believe myself so far north,' Murchison wrote. 'Fine meadows, rich vegetation, a noble river, and no mountains; and yet I am now four degrees north of the Orkneys.'[12] And he commented, 'I am most agreeably surprised in finding the condition of the Russian peasant so infinitely superior to what I had anticipated. They are, in general, well housed and decently clad; so infinitely superior to the Irish and even many Londoners, that no sort of parallel can be drawn. Beggars are almost unknown.'[13]

But the party's progress was slow. If the people were well housed, the rest of the infrastructure was dreadful. The road frequently degenerated into 'deep, fathomless sand without any attempt at foundation'[14] and the horses quickly tired. The Russian authorities operated a post-house system, where exhausted animals could be swapped for fresh ones. But conditions in the country were so tough, and the mortality rate among horses so high, that the post-houses routinely ran out of animals; one post-house master told Murchison he had recently lost forty of his sixty horses. To make up for lost time, when the track was firm Murchison harnessed five or six horses to each of his carriages and galloped down interminable forest tracks 'over sand, boulders and bogs, at the rate of often as much as ten or twelve miles an hour ... whirling over a broken wooden bridge with the noise of thunder'.[15]

The settlements grew sparser as they drove further north,

and the people more suspicious. The Russian administration had a well-earned reputation for repression and corruption, and the villagers automatically assumed the party had come to extract further taxes. Von Keyserling worked hard to disabuse them, but the local population remained profoundly reluctant to offer information about anything that might be taxed, including the presence of local quarries.[16] Although Murchison could see all around him signs of limestone extraction in the form of lime-washed buildings, the local guides in this strange land were unwilling to assist him.

It took the party another two and a half weeks of hard riding to reach the shores of the White Sea and the Arctic trading port of Archangel. After the excitement of the Volkhov cliffs, they had found little new of geological interest, and Archangel was another disappointment.

Pictures from the time show wide, muddy streets lined with dilapidated wooden houses, a bleak waterfront dominated by a government building and a skyline punctuated by the cupolas of churches. Out in the bay a scattering of ships lay at anchor, their sails furled, and small rowing boats idly ferried men back and forth. Archangel was being supplanted by St Petersburg as Russia's main trading port with Europe, and the town had hit hard times. 'One long death-stricken street, with wooden houses at short distances, is the main feature,' Murchison wrote with flagging spirits, after checking into a 'stinking wretched house near the river.'[17] But to their surprise, riding along the waterfront, the party encountered a crew of British sailors loading timber for Newcastle and were able to negotiate the repatriation of their growing collection of souvenirs, including a pickled seal's head, animal skins and ivory carvings made by the local Samoyedic people.

Over the next few days Murchison and de Verneuil mounted a series of field trips along the thickly wooded shores of the

White Sea. To the east of the city, in low cliffs along the water's edge, they were cheered to find grey horizontal bands of Carboniferous Limestone sitting conformably on top of beds of older Devonian sandstone: a clear illustration of the upper boundary of the Devonian. Murchison could now categorically confirm the stratigraphic position of the Devonian as lying between the Silurian and Carboniferous periods. That, at least, was something.

Archangel was the furthest north the expedition had permission to travel, and in early July the party turned south again, heading for a range of sparsely populated hills that separate the catchment area of the White Sea from the headwaters of the River Volga and its tributaries.

They followed sandy tracks upstream beside fast-flowing Arctic rivers swollen by melting snow. Sometimes the riverbed was up to half a mile wide: in other places it had broken up into a mass of braids that meandered across a boulder-strewn valley floor. All around, the hills were densely wooded with birch and pine. Time and again their heavily laden vehicles sank axle-deep into a quagmire of sand and mud and had to be dug out. As they reached the upper waters of one valley they climbed over the wooded watershed into the next, with range after range of blue-green hills disappearing into the distance.

Moving south, they passed small cliffs where Murchison could make out an underlying bedrock of colourful beds of 'mottled red and yellow' sand and limestone,[18] which reminded him of a puzzling reddish-brown rock he had seen in north-east England. There it was known as Magnesian Limestone because of its unusually high magnesium content. Nobody had paid much attention to it, and it occupied an uncertain position in the stratigraphic table between the coal seams of the Carboniferous and the still younger rocks of the Triassic New Red

Sandstone geological period (see figure 1.1, p. 10): a boundary that had never been properly determined. To add to Murchison's confusion, he found the fossil remains of an oyster-like bivalve in sandstone that he had never seen before.[19] 'I entered into a long cogitation on the probable age of these red sandstones,' he wrote in his journal.[20] Should he classify them as late Carboniferous, early New Red Sandstone or perhaps an entirely distinct geological period that lay between the two? He was accustomed in Britain to seeking the advice of local experts, but there was nobody in Russia to ask. Murchison was on his own.

The group continued south, past remote golden-domed monasteries and small logging towns, the river clogged with rafts of newly felled trees waiting to be floated downstream to the sawmill. At a garrison town designated by the authorities as a place of internal exile for political dissidents, the visitors were greeted by the vice-governor, who insisted on escorting them to the officers' club and entertained them with whist and billiards.[21] Once again Murchison was impressed by the Russian character. 'The people seemed to be very poor though well mannered and civil and a fine young strapping peasant, who lighted a piece of swam or tinder from my cigar, fell at my feet and kissed my boots because I gave him a piece of 5 copecks in silver,' he wrote in his journal.[22]

Finally, after more than a month on the road the party reached the swampy headwaters of the Volga. As they travelled downstream they frequently sank deep in the mud and were forced to cross the river in search of firmer ground. 'On our arrival at one quay,' recorded Murchison after one such crossing, 'we found the men more than half drunk', and the dismantling and loading of the coaches onto impromptu ferries was accompanied by a chaotic 'din of many voices and the want of control'.[23]

The river grew steadily wider and the valley more populous.

Birch and pine forests gave way to water meadows and open fields. Church towers and onion domes appeared on the skyline. They paused in the popular religious centre of Yaroslavl, bustling with pilgrims, where Murchison noted with delight that he had passed his first night in weeks 'with not a bug in my room'.[24]

Downstream from Yaroslavl, the party examined more exposures of the puzzling red-and-yellow rock laid bare in the cliffs by the fast-flowing water. There were also steep gullies of milky-white gypsum cascading down the flanks of the hills like frozen waterfalls. Finally, crossing slopes of apple orchards ripening with fruit, they arrived at the ancient trading port of Nizhny Novgorod, strategically sited on a series of promontories overlooking the river.[25]

It was August. The annual summer fair was in full swing: a forest of ships' masts lined the river bank and hundreds of storekeepers had decamped to a tongue of flat land beside the water, transforming it into a noisy market. The cries of hawkers filled the air, and wisps of blue smoke from hundreds of fires drifted upwards, carrying with them the sharp tang of grilling meat. '200,000 persons are daily present for upwards of a month,' Murchison noted with interest, 'and including the comers and goers, not less than 350,000 visit it, there being many representatives from all the Eastern and Tatar countries of the Czar.'[26] It was the women, however, who particularly caught his attention: 'we were now fairly among picturesque female peasants. The sombre and heavy dresses of the north were now exchanged for bright red helmet-shaped caps and short jacket-like gowns fringed with red. Some of the girls wore their hair hanging in long ringlets, and they had long white stockings upon their stout and brawny legs.'[27]

Nizhny Novgorod was their furthest point east and, after revelling in the pleasures of big-city life, the party turned west, following another series of river valleys towards Moscow. The red-and-yellow sand and limestone outcrops gave way to beds of older grey Carboniferous Limestone replete with 'true Carboniferous Limestone fossils': the group was moving back through geological time.[28] They paused in the country's former capital of Moscow and then took the new highway – the best road in Russia – to St Petersburg. They had been travelling for two and a half months and had covered around 4,500 miles. On the steamer home, 'dropping down the Baltic', Murchison told Sedgwick he was in 'very good humour'.[29]

Back in London, he took stock. There was no question in Murchison's mind that the expedition's highlight had been the unequivocal establishment of a link between the Old Red Sandstone and the Devonian. 'The mixture of Devonian shells of Devonshire with fishes of the Highlands of Scotland demonstrates that Sedgwick and myself were right in identifying the Scottish Old Red with the Devonshire rocks under the name of Devonian,' Murchison rejoiced.[30] His advocacy of the Devonian as an important and distinct period in the Earth's early history between the Silurian and Carboniferous systems was looking ever more justified. But the trip had also thrown up some interesting new questions, including the nature of the mysterious multicoloured sand, mud and limestone cliffs that he had repeatedly encountered.

Murchison was still puzzling over these curious rocks when he received an unexpected invitation to return to Russia. It came from the Russian government and asked if he would be willing to extend his mapping of the country's geology. The Russian authorities had woken to the fact that they were sitting on rich coal reserves that could be used to drag the empire into the industrial age. But there was a profound lack of detailed information.

It was an attractive proposal. Murchison regarded his stratigraphic work on the Silurian System in Britain as more or less complete; another trip to Russia would give him the opportunity to further substantiate the universal applicability of the Devonian System and, more importantly, spread his ever-expanding 'kingdom' of 'Siluria' to new corners of the globe. It would also significantly enhance his reputation as a geologist of international stature. 'The carrying out of this plan, and making of a true and good geological map of Russia in Europe,' he mused, 'would be a work on which I might be content to rest my claims to a name among practical geologists.'[31]

As a result, by April 1841 he was back amid the 'cupolas and bright white buildings' of St Petersburg, where he received a gratifyingly ceremonial welcome from the tsar, Nicholas I. Murchison was impressed. 'When the Emperor is in full costume it is impossible to behold a finer sample of human nature,' he wrote.[32] 'The tight breeches are unusually well fitted, especially on the upper part of the thigh, so as to delineate even the virile member with great precision. Your eye glances from these to his beautiful family, and the animated but épuisée Empress, and the history is told.'[33]

The geologist brought with him rough-and-ready maps of the country's geology compiled by previous visitors, and explained to the tsar that he was interested in 'the structure of a country the whole northern region of which was made up of strata which I had spent so many years in classifying and arranging in other parts of Europe; how their vast scale in Russia had surprised me, and how they offered evidences which were wanting in the Western countries'. They also talked about coal: Murchison explained that his travels in the north of the country had shown the area to be almost entirely made up of Silurian and Devonian rocks and therefore too old to contain any, but

6.1 Murchison's mighty journey through Russia in 1841.

'I then comforted him about the great coal-field of the Donetz, in Southern Russia, to which I was destined to go.'[34] The tsar was intrigued and offered Murchison every assistance, including 'a formidable Imperial document, countersigned and double-sealed to enforce attention from all persons in authority along their route'.[35]

Murchison was once again accompanied by de Verneuil and von Keyserling. The plan was to travel east across 'European' Russia as far as the Ural Mountains, then head south to the borders of Kazakhstan and the Caspian Sea and finally return north via the newly emerging Ukrainian coal-mining region of the Donbass. Spring had arrived, and as the group retraced its steps through Moscow and east along the valley of the Volga, birch trees were bursting into delicate green leaf. In the villages, cherry orchards were heavy with white blossom.

Sometimes they continued travelling through the night, coated in the dust thrown up by their carriages. They passed 'gaudy, often dirty, rickety, tumbling Russian towns', their 'planks all awry',[36] and low cliffs of 'interminable red sandstones and marls',[37] familiar from the year before, but there was a disheartening lack of fossils to help identify them. The region has no 'back bone', Murchison wrote in exasperation, and 'malevolent critics would say the people of the surface stuff reflect the substrata – slippery and treacherous'.[38] And he wrote in bad doggerel:

> How bored I am with the red sandstone,
> I would certainly pay a thousand Louis
> To never again in my life see
> The new red sandstone of the Tartare.[39]

The small group passed through Kazan, where the slender towers of the kremlin dominated the skyline and they were entertained by 'the fat jolly Vice-Governor', and then continued east.[40] The Volga here turns south towards the Caspian Sea, and the party left the river to follow the wide, shallow valley of the River Kama east towards the Urals and Siberia.

Stands of silver birch lined the banks, and beyond them the green hills of central Russia put Murchison in mind of the Welsh borders. In places the river stretched a mile wide. The weather grew warmer, and clouds of mosquitoes hung menacingly in the air like 'steaming bands of quivering light, each layer of which was made up of these countless beings'.[41] Temperatures in the shade climbed to thirty degrees Centigrade. Safe drinking water became increasingly hard to find. 'Our thirst was great,' recorded Murchison at one point. 'I ate nothing yesterday or today but bread and two eggs ... [and] a little cool quass [a mildly alcoholic drink made from cereals and stale bread], being unable to touch the water.'[42]

Eventually, after days of floundering along the marshy banks of the river, they climbed out of the Kama Valley and hit the great 'Siberian Road', one of the few well-trodden trading routes running east towards the Urals. In the distance the tower of the great white church of Perm 'burst into view'.[43] And as the small group paused in the newly emerging industrial city, they ran across a colony of political exiles condemned to work in the city's primitive copper mines and iron foundries. Murchison's notes once again betrayed little sympathy for the exiles, but he observed that the city 'smells of Siberia'.[44]

They continued east, making a series of gentle ascents towards the long, low ridges of the Urals. Behind him lay 'the wooded mass which separated us from, and shut out all sight of, Europe'; in front 'the vast wilds of Siberia, rolling over and over in countless masses of eternal fir and pine'.[45] Nearby Murchison may have noticed a famous stone column – the western face bearing the word 'Europe', the eastern one the word 'Asia'. But the geology hadn't changed much: along the track they found more outcrops of the colourful bands of sand and limestone, or the 'Red System', as Murchison was now calling it.

Finally in late June the party arrived, hot, dusty and weary, among the 'white and green churches' of the gold-mining centre of Ekaterinburg, sitting astride the highway between Europe and Siberia. The town proved unexpectedly elegant, with churches,

6.2 'Peeking into Siberia' from the heights of the Ural Mountains. Reproduced by permission of the British Geological Survey CP20/076.

fine towers and public buildings lining the main streets. After weeks of cramped, vermin-infested village lodgings, Murchison was delighted by the luxury of city living. 'Here I am, writing in one of the best and cleanest [inns] which can be met with,' he wrote of his enormous suite of rooms in the city centre. 'My sitting room is furnished with green sofas round it, a well painted floor and not a curtain or shutter to exclude the light of six staring windows. Here I receive my visitors. My bedroom, through which they pass to reach me, contains my little traveling iron bed, superadded to the usual tables and chairs.'[46]

Murchison had been told that the Urals contained vast stretches of Silurian rock, and over the following weeks he explored the wooded mountain slopes in the hope of claiming yet more territory for his own empire. De Verneuil and von Keyserling found bands of Silurian, Devonian and Carboniferous sandstone and limestone in the river valleys, flanked by ridges of igneous rock. Many of the streams had been diverted to provide power for the mines, and in the dried-up beds there were clear cross-sections through the strata.

Murchison meanwhile visited copper, iron, gold and diamond mines. The region, he wrote, appeared to have been 'washed by a general shower of gold'.[47] From time to time he encountered the regular shuffling caravans of convicts heading for penal colonies in the far east:

> the men, chained two and two, stepping out stoutly to the beat of the drum, glad to reach the halting place at 10 o'clock, and before the great heats of the day ... Twenty or more little carts followed with the aged, sick and provisions as well as the wives of those who were allowed such a luxury ... Thank God, in England, we have the sea for our high road to banishment, for such scenes are very harassing.[48]

But he was also delighted to find significant outcrops of Silurian rock that contained fossils – 'I could scarcely distinguish from my own *Pentamerus knightii* of Aymestry and Ludlow'[49] – although he bemoaned the lack of accurate maps to record them on. 'Were I Emperor of Russia I would make, verily, at least one thousand of my lazy officers work for their laced coats, and produce me a good map,' he confided to his journal.

> I never shall forget my surprise last year at Nizhny Novgorod, when the Government House was ran-sacked for a map, upon which my line of march to the south of Moscow was to be traced ... [and] at length what came forth from this centre of Russian wealth and commerce [was] one of the little three-rouble maps which the common traveller buys, with simply the names of the chief places and small towns.[50]

Even so, by the end of June he was content to have arrived at a broad understanding of the geology of a great stretch of western Russia. A huge plug of his 'Red System' spread across the entire centre of the region, resting in a bowl of older Carboniferous, Devonian and Silurian rocks that rose to the east and west. There was no mention of Sedgwick's even older Cambrian strata. But the question of how to categorise these 'red rocks' remained. By now Murchison had broken them down into three bands: the top two were strikingly similar to Triassic New Red Sandstones that he had seen in Britain, while the bottom band contained large tracts of what looked like the mysterious Magnesian Limestone he had met the year before. The question was whether to bracket them with the Carboniferous rocks below or the Triassic Period above. Or could they represent an entirely new and hitherto unrecognised geological period in the Earth's history? Murchison reserved judgement on this last possibility. By early July the survey of the Urals more or less complete – it was time to move on.

The party picked up the River Ural[51] and followed it south towards Kazakhstan. The banks became 'wide, monotonous and featureless': trees and mud gradually gave way to 'bare, barren and bad' grasslands.[52] They were entering the sandy plains of the Russian steppe, with grasslands stretching as far as the eye could see. 'This day the British Association is assembling at Plymouth and I drank success to it,' wrote Murchison in his journal. 'How few of the members there will have lighter hearts than their general secretary in Siberia!'[53]

They met camps of Kirghiz people, trading leather and cloth, and herds of free-roaming camels, dromedaries, goats and oxen. Dried dung replaced wood as a source of fuel. Clouds of starlings blackened the sky, and flocks of curlews lifted off as they passed. Hawks hung in the air above them.

At the Kazakh border, little more than a dusty line in the sand, they turned west, driving for hundreds of miles along sandy tracks through featureless grassland burnt yellow by the summer sun. Finally they reached the banks of the middle reaches of the Volga, 'burnt up, arid and without grass', a very different river from the one they had followed further north.[54] They passed low river cliffs where chalk sat on top of 'shaley, gritty, thin beds' of limestone, suggesting that the geology was changing once again and they were entering a new and younger geology, which Murchison would later identify as the Jurassic Limestone and Cretaceous chalk of the lower Volga Valley.

Further south, the grassland thinned out still further and was replaced by sandy desert. 'The prospects of a famine increase at every step,' Murchison wrote.[55] Temperatures rose once again, and a dusty, treeless expanse reached to the horizon. They hit deposits of mud and sandstone containing the fossil remains of bivalves such as Lucina and Nucula and the sea-snail Turritella, signs that the rocks were growing still younger and they had moved out of the Cretaceous into Tertiary muds. It was now clear to Murchison that just as the central region of the country was dominated by rocks of the 'Red System' and the Devonian and Silurian periods, much of this southern region of the country clearly post-dated the Carboniferous and he could safely report to the tsar that there was unlikely to be coal anywhere across the southern steppe.

By mid-August they had reached the port of Saratov on the lower Volga, home to a colony of Volga Germans invited a hundred years earlier by Catherine the Great to populate the region and improve its agriculture.[56] But now the weather changed and the notorious late-summer rains of southern Russia set in. The mud season had begun and the roads turned to quagmires. 'The Russians know this autumnal time well, and every

6.3 The Russian steppe with Orenburg on the horizon. Reproduced by permission of the British Geological Survey CP20/076.

sensible man has already struck work,' Murchison wrote with a sinking heart. 'We only are plunging into the mire!'[57]

They passed hurriedly through the small, wet village of Tsaritsyn, an 'abandoned and wretched fortress',[58] which would later blossom into the huge city of Volgograd, and in late August finally bade farewell to the arid valley of the Volga. Time to strike west, across a sandy plateau towards the Ukrainian region known as the Donbass, an area of scattered Cossack villages and scrubby corn fields that took its name from the River Donets.

They were back in the older rocks of the Carboniferous. Thick seams of coal had recently been discovered close to the surface, and a small coal-mining industry was emerging. Murchison had promised the Russian authorities he would investigate its potential. He took an almost instant dislike to it. Fortune-hunters and adventurers from across Russia had moved in, lured by the promise of coal and quick money, and the region had gained a reputation for violence and lawlessness. In the towns the streets were clogged with wagons and camels. Murchison found most of the mines to be small and grossly inefficient. At one, the tunnel had been driven horizontally into the hillside without pit props or roofing timbers and frequently collapsed. Another was operated by a 'military colony' – a peculiar Russian institution in which workers divided their time between mining and military duties – and the main shaft had lain flooded for two years.

'The system here is more absurd than can be imagined,' Murchison fulminated. 'The country (and pits of course) belong to the military colony which composes the village. These are ruled by their own bureau and the Ministry of War. They ask for an engineer of mines, and the unfortunate man who conducted us through the works was sent ... [and] here he remains with the mines full of water.'[59] To Murchison, it was clear that the engineer had no interest in what he was doing. The Donbass, he concluded, was 'a nasty country', its people 'disposed to thieve'.[60]

Ten days later he had had enough. It was time to return to the tsar. 'The road to Moscow, after this hot and parching summer, resembled one which has been passed over by a retreating army,' he wrote in horror. 'There were so many dead horses, that we counted seven or eight on the roadside, in 30 versts [20 miles], on the carcases of which the magpies and crows were feasting.'[61] Gradually they passed through the Russian towns of Kharkov, Belgorod and Kursk and the geology changed again: they were travelling back in time, from the grey rocks of the Carboniferous to the sand, chalk and 'rusty yellow' grits of the Old Red Sandstone or Devonian, last seen in the Urals.

Murchison spent a week in Moscow preparing his report and then pushed on to St Petersburg, where his audience with Nicholas once again warmed his heart. 'The moment the door opened, his radiant, noble countenance, and his forward hand, assured us of the kindest reception,' Murchison reported in the awed prose that he reserved for such occasions. 'If I felt a warm pressure of his powerful grasp in the Spring, it now appeared to me that his shake of my hand was that of a kind friend who welcomed my return.'[62] They discussed the expedition and Russia's probable gold and coal reserves. Murchison reiterated his old argument that coal was never found in Greywacke rocks, and confirmed his belief that the country's biggest coal reserves were in the 'great productive Carboniferous tract of the south' in the Donbass.[63]

Finally, in late October, Murchison headed for London. He had spent five months in a horse-drawn carriage covering a distance roughly equivalent to the diameter of the Earth.[64] He was sick and weary. At night he experienced bouts of delirium. The doctor at the British Embassy advised him not to travel. But Murchison was determined to get home before the winter storms hit the Baltic and North seas. On 1 November 1841, despite heavy fog, he entered the mouth of the Humber, en route for London.

Over the next nine months, sitting in his study in Belgravia, Murchison digested his notes and correlated the fossils he had collected. He had successfully expanded 'his' Silurian System to cover large tracts of Eastern Europe and Russia and had further firmed up the existence of the Devonian. But it was Russia's 'Red System' of rocks that had been the real surprise.

In April 1842 Murchison and de Verneuil told a meeting of the Geological Society in London that the mysterious reddish limestone and sandstone rocks they had found across large areas of central Russia clearly resembled, in look and composition, the Magnesian Limestone outcrops in northern England and parts of Germany. Far more importantly, the fossil record suggested a flora and fauna intermediate in character between the

Carboniferous and Triassic geological periods. One creature in particular, a microscopic crustacean resembling a mussel and known as Cytherinae, had never been found in Carboniferous or later Triassic New Red rocks and therefore pointed towards an entirely unfamiliar ecosystem: a hitherto unrecognised period in the Earth's history sandwiched between the Carboniferous coal-bearing rocks and the Triassic sandstones. Murchison and de Verneuil called it 'the Permian', after the Russian city of Perm.[65]

BASIC CATEGORY	GEOLOGICAL PERIOD
SECONDARY ROCKS	Cretaceous
	Jurassic
	Triassic
	PERMIAN
	Carboniferous
TRANSITIONAL/GREYWACKE	Devonian
	Silurian
	Cambrian
PRIMARY ROCKS	

6.4 The new stratigraphic table, after Murchison's discovery of the Permian in 1841.

Given the struggle to win acknowledgement for some geological periods, it is striking that the Permian was received by geologists across the world almost without question and rapidly became an accepted segment of the stratigraphic table of the Earth's rocks.

Murchison's Russian journeys of 1840 and 1841 are regarded as among the great scientific expeditions of the nineteenth century, comparable to those to South America, the Himalayas and the source of the Nile.[66] They turned Murchison from a run-of-the-mill British geologist into an international figure. To many educated Victorians, his achievements expressed the power and significance of the British Empire. The British Parliament had just laid claim to New Zealand; the British East India Company governed much of the Indian subcontinent; and David Livingstone was preparing to lay claim to vast tracts of the African continent. Now British geology provided the model that enabled the identification and classification of rocks across the world. John Bull was lording it not only over European geology, but across the globe.[67]

Murchison shared many of these domineering attitudes. He was now referring quite unselfconsciously to 'my' Silurian System, and had come to regard it as a private fiefdom. It might be compared, observed his biographer Archibald Geikie, 'to a vast business which he had established, of every detail of which he was complete master, and which he laboured to extend into other countries'.[68]

Not everybody was happy with the situation. The historian of geology James Secord quotes an 'old Indian hand' reflecting on years of experience in the Himalayas and wondering why his thinking should be 'changed according to Mr Murchison's last paper on Shropshire'.[69]

Unassailable as it now appeared, however, the foundations of Murchison's empire were about to crack.

Chapter Seven

SILURIA V. CAMBRIA, 1841–52

Near the centre of Cambridge, behind the fancy shops, multi-storey car parks and cheap food outlets, lies the University's Sedgwick Museum: home to one of the largest collections of fossils in Britain. Here the specimens are stacked high in dark wooden display cases, with exhibition notes often written in neat sloping handwriting. It's a gem of a place, where they expect you to want to know the full details of each fossil: genus, species, location found and the rock type in which it was discovered. The museum is said to contain two million specimens, ranging from 500-million-year-old Cambrian trilobites and brachiopods through to Silurian bivalves, Devonian fish and a large collection of Jurassic ammonites and belemnites (or 'thunderbolts').

Sedgwick wouldn't recognise the place. When he inherited the University's fossil collection in 1818, it consisted of several boxes of specimens collected mostly by the eighteenth-century surgeon and naturalist John Woodward.[1] Over the years Sedgwick had added to it, sometimes contributing his own funds, and by the late 1830s it had grown to some forty boxes and crates, containing more than 100,000 specimens, stored haphazardly in his rooms.[2] Many had never been unpacked or identified, and Sedgwick was approaching a state of what he called, with considerable understatement, 'geological congestion'.[3] Finally in 1841, after a wait of more than ten years, the collection was offered a permanent site at the University's new library in the so-called Cockerell Building.[4] As Murchison prepared for his second trip to Russia, Sedgwick was finally able to begin unpacking his expedition boxes and discover what he had.

There was a growing urgency. The success of Murchison's book on the Silurian System, and its apparently solid foundation in the fossil record, made claims for a distinct Cambrian System look increasingly weak and uncertain. Many geologists were finding Sedgwick's pick-'n'-mix reliance on structural, lithographic and fossil markers too complicated and idiosyncratic to apply. His credibility wasn't helped by a continuing failure to produce his own long-awaited 'big book'. And so, for many geologists, the legitimacy of the Cambrian System was still up for debate. Charles Lyell spoke for many when, in a new edition of his influential *Elements of Geology* published in the summer of 1841, he asked whether the Cambrian could justifiably be maintained as a distinct geological period.

Sedgwick set to work with uncharacteristic doggedness on the 'enormous labour of unpacking'. Over a period of three months, at 'great cost to the University and great to myself', he emptied box after box.[5] Crinoids, brachiopods, trilobites and coral shells were spread across the floor and benches as he began the search for evidence of a uniquely Cambrian form of life. The specimens came from all over north and central Wales; many were broken, fragmented and hard to identify. But as Sedgwick worked through them, placing each one in its appropriate family group, one thing became abundantly clear: none of them were unique to the Cambrian rocks. They had all been seen in Lower Silurian Caradoc Sandstones as well. Indeed, Murchison had used many of them to illustrate his book on the Silurian System.

Finally, in the autumn of 1841 Sedgwick announced that he had come up with an explanation for his apparent failure.

He would reveal it at the November meeting of the Geological Society. As members gathered in the Society's grand premises overlooking the Thames on the Strand,[6] there was an air of expectation: how had he squared the circle? Sedgwick began by telling the packed auditorium that he accepted there seemed to be 'very few new types of creation'[7] in the Cambrian Period. But that didn't reduce its importance; what it signified was that during this period in the Earth's history new life forms had begun to appear that would only reach their full expression in the later, Silurian Period. The two systems shared the same life forms: what made them distinct was that as new life appeared, the number of different creatures (and the predominance of particular species) changed through time. In other words, the Cambrian Period was marked by the gradual and progressive appearance of what would later become Silurian life, rather than by its own unique fossil record.

This was not an entirely new concept. Five years earlier Sedgwick's former companion Charles Darwin had returned from South America and the Galapagos bringing with him the revolutionary theory of evolution: the notion that life forms changed and developed over time. Sedgwick, like many devout Christians, was unable to accept anything so radical as evolution, but he did acknowledge that life had changed over time, in accordance with some vaguely perceived notion of God's ever-changing plan for the Earth: like a scroll, as the historian of geology James Secord has described it, gradually unrolling and revealing ever newer forms of life. Indeed, this concept of an unrolling scroll was the basis on which the Devonian was first accepted as a separate period in the Earth's history: it was when life had moved gradually from the sea to the land.

Drawing on the same logic, it was therefore entirely possible that 'Silurian' organisms might also make a gradual appearance, so that there was no unique set of Cambrian fossils. Yet the suggestion that there was not a single life form that was unique to that period was new, and Sedgwick's fervent advocacy of a so-called 'progressionist' model failed to win the support of his colleagues.[8] Many in the audience had hoped for a simple description of a unique fossil fingerprint, and the general reaction was one of 'disappointment and dismay'.[9] The Irish physician and geologist William Fitton, a former president of the Geological Society, summed up the feeling of many members when he argued that until the Cambrian Period could be characterised by discrete organic remains, its separation from the Silurian System was merely a matter of convenience.[10] Edward Bunbury, a young barrister in the audience, later reported to an already doubtful Charles Lyell:

> as the paper proceeded & we heard of Caradoc sandstone fossils in the heart of Snowdonia & first Cumberland & then north Wales seemed to give up as undistinguishable from the lower Silurian system, every-one was disposed to cry out 'The Cambrian System, where is it?' And the Echo answered Where? – from the beginning to the end it was not to be found.[11]

The winter and spring of 1841–2 must have been a difficult time for Sedgwick. Murchison had returned from Russia with a new confidence and now regarded himself as an international figure working on a world canvas. He could claim to have mapped the geology of large areas of Europe and Russia and to have identified, named, and won international acceptance for, three new geological periods: more than any other geologist could boast of. Moreover, his wife Charlotte had recently come into a large inheritance and the couple had moved from their relatively modest home in London's Bryanston Place to a much grander establishment in Belgrave Square, in one of the capital's most fashionable districts.

Murchison now regularly hosted grand parties and soirées. 'Ministers of State, men of rank, of science, of literature, of art, and of travel mingled there together,' according to one account.[12] 'I don't know what fortune Murchison has, but this is coming it rather strong,' wrote a young John Ruskin, then an up-and-coming artist and critic, of one party. 'Rooms all pale grey and gold – magnificent cornices with arabesques like those of Pompeii in colour, furniture all dark crimson damask silk & gold – no wood visible – at least four footmen playing shuttlecock with people's names up the stairs.'[13] Murchison's past failings, his less-than-brilliant military career and his failure to support himself as a country gentleman had melted away. It is said that he now deliberately 'cultivated the character of a retired officer' and, in the course of countless after-dinner speeches, 'never tired of relating his early exploits in the Peninsular War'.[14]

In all this he was ably backed by Charlotte, who had amassed 'a fine fossil collection of her own' and was widely respected as 'cultivated, intelligent and able to converse on a wide range of subjects'.[15] Friends noticed that Murchison was becoming grander and more self-important by the year and, in the words of one colleague, they could trace in him 'an increasing impatience of opposition and a growing tendency to discover in the work of other fellow-labourers a want of due recognition on their part of what had been done by him'.[16] He was also prone to 'speaking of the researches of his contemporaries, specially of younger men, in a sort of patronizing or condescending way,' observed another. 'He had hitherto been, as it were, one of the captains of a regiment; he now felt himself entitled to assume the authority of a general of division.'[17]

The contrast with Sedgwick's more humdrum provincial life was painfully obvious. Sedgwick now occupied large and comfortable rooms overlooking the picturesque Great Court at Trinity College. There was a bedroom; a dining room lined with watercolours of the Yorkshire Dales, Norwich and the Lake District and furnished with plenty of chairs and sofas; and a second book-lined room where he mostly sat, 'the table and most of the chairs covered with letters and papers'.[18] He also gave and attended parties, constantly appearing 'without special invitation at the tea table of his intimate friends', where his reputation as a raconteur meant that 'when he started on a story there was a dead silence until he had finished'.[19] But these were small affairs compared with Murchison's glittering soirées, and Sedgwick was still hemmed in by Church and university obligations.

He continued to spend the winter months on prebendary duties at Norwich Cathedral, where life followed a monotonous routine. 'I rise early (often between five and six) and do all my work in the mornings, often before anyone is stirring,' he wrote to a friend. 'At nine I meet my servants – quarter past, breakfast – at ten morning service at the Cathedral – after service odds and ends – callings and shopping and I know not what – lunch at one – then a scamper on horse-back (when I have time) with my niece ... Cathedral service again at four – dinner at six.'[20] Back in Cambridge, he faced an equally burdensome round of obligations. 'I wish you could see me, looking sour as a crab, and dignified as a round of beef surrounded by sixty-nine unfortunate undergraduates, whom I am tormenting with long Latin passages that I have given them to translate,'[21] he wrote to his niece Fanny, one of his many young female correspondents.

Meanwhile his health was worsening. Cambridge is a low-lying, damp and cold city: sharp, icy winds from Scandinavia and northern Russia barrelled down Trinity Street, and winter after winter Sedgwick was struck down by bouts of flu and gout, which left him prostrate and unable to work. 'During three months ... I was almost confined to my fireside (and part of the time to my bedroom) by an attack of influenza,' he wrote to Kate Malcolm, the daughter of an old friend and another of his regular female correspondents.[22] 'As soon as the influenza left me, the rheumatic gout took its place; so I went and tried to boil the fiend out of my bones in the hot springs of Bath where I remained during the month of April.'[23]

More importantly, he had made slow progress with his 'big book' on the older Greywacke. 'It was conceived on too large a scale and conducted with too much minuteness of detail, to be completed in a life-time so much taken up by other occupations and so harassed by continuous and increasing ill health,' wrote his biographers, John Willis Clark and Thomas McKenny Hughes.[24] Sedgwick himself, in his more reflective moments, was aware of the problem. 'Perhaps I may retire for a year into Germany, and hire lodgings at some of the watering-places,' he wrote ruminatively to Murchison. 'Then I could easily make a smash of my volume. When I am well I can write fast enough. And it seems to me now that all my materials are well in hand. But in Cambridge I am bothered out of all my senses.'[25]

Murchison was supportive, but by now it was clear that he had his own agenda. In February 1842 he was elected president of the Geological Society for the second time, and in his presidential address he returned to the problems of the boundary between the Cambrian and Silurian Systems. 'I put a deal of powder and shot into it, foreign and domestic,' he told Sedgwick.[26] It was 'a laboured production', according to one listener, 'occupying forty of the closely printed pages of the Society's Proceedings and must have somewhat exhausted both reader and audience from its mere length'.[27] It was also unapologetically self-congratulatory. 'The perpetuity of a name affixed to any group of rocks through his original research is the highest distinction to which a working geologist can aspire,'[28] Murchison claimed in somewhat bombastic prose, presumably referring to his own naming of the Silurian, Devonian and Permian periods.

But the meat of his address concentrated on an idea that he and others had already entertained. It remained to be shown, he argued, 'whether there [are] any type of fossils in the mass of the Cambrian rocks different from that of the Lower Silurian series. If the appeal to nature should be answered in the negative, then it [is] clear that the Lower Silurian type must be considered the true base' of the Greywacke rocks and the Cambrian should be confined to the very oldest rocks, which contain no fossil record.[29] The 1834 fudge, which eight years earlier had arbitrarily assigned the 'Bala' rocks at sites like the quarry at Gelli-grin to the Cambrian, and the strikingly similar 'Caradoc' rocks at the small Shropshire village of Meifod to the Silurian, had come back to haunt Sedgwick. Murchison was now claiming the Bala Limestone for his own, in what would become known as the extended or expanded Silurian System,[30] a concept that would not only increase the physical size and spread of his fiefdom across almost all of Wales, but would also mean that Siluria marked the moment in the Earth's history when 'organic existence' had begun.[31]

Sedgwick missed the meeting. He must have read the subsequent transcripts with growing dismay. If he was to ward off Murchison's attempts to grab the Bala Limestone and all the rocks around it, then he needed evidence to support his 'progressionist' ideas.

*

THE 1834 AGREEMENT		THE EXPANDED SILURIAN SYSTEM	
UPPER SILURIAN	Ludlow Rocks	UPPER SILURIAN	Ludlow Rocks
	Wenlock Limestone		Wenlock Limestone
LOWER SILURIAN	Caradoc Sandstones	LOWER SILURIAN	Caradoc Sandstones Gelli-grin/Bala Limestones
	Llandeilo Flags		
UPPER CAMBRIAN	Berwyn Slates		
	Gelli-grin/Bala Limestones		Llandeilo Flags
LOWER CAMBRIAN	Snowdonia Slates	CAMBRIAN	

7.1 Murchison's 1842 expanded Silurian System.

Six months later, in the summer of 1842, Adam Sedgwick returned to Wales with a new plan of attack. During earlier visits he had noticed that, in addition to the bands of Bala Limestone running through the Berwyn Hills, there appeared to be evidence of additional limestone bands running roughly north–south through the hills of the region, like the marbling on a piece of steak. He had always assumed that since the rocks of north Wales appeared to grow older as you travelled from the east to the west of the country, the limestone strata would follow the same pattern. His new idea was to return to these faint bands of limestone – one of the few sources of fossils in the region – in the hope that as he travelled across the open moors from Snowdonia to the English border he would stumble on fossils that supported his 'progressionist' ideas.

He already had tantalising evidence. Some months earlier Sedgwick had received a series of letters from an 'amateur' geologist called John Bowman, a retired bank manager from Wrexham who made enough money by the age of forty-five to devote himself full-time to natural history. Bowman had spent the last two summers scouring these limestone bands: in one of the most westerly outcrops, on the northern slopes of Snowdonia, he had found Lower Silurian fossils, including the trilobite *Asaphus buchii*, regarded as a key marker of Murchison's Lower Silurian Llandeilo Flags.

Yet Snowdonia was deep in an area regarded as Cambrian. Bowman's findings pointed to one of two conclusions. Either the Silurian rocks in Wales were far more extensive than anybody had so far imagined, in which case the Cambrian Period rested on even weaker foundations, or – Sedgwick must fervently have hoped this was the case – it was possible proof of his 'progressionist' model.

Could it be that this fossil was an isolated representative, an early arrival of the species *Asaphus buchii* during the Cambrian, before it reached its full flowering in the later Lower Silurian? If so, then creatures like *Asaphus buchii* should become increasingly common in the limestone bands as you moved east towards the younger rocks of the English border.

Sedgwick had hoped to make the trip with Bowman himself, but that winter the retired banker died unexpectedly. In his absence, Sedgwick took with him one of the country's leading palaeontologists, a 'nervous excitable character'[32] named John Salter who, at the age of twenty-two, had won widespread respect for his expertise – an indication perhaps of how recent the emerging study of fossils still was. Salter had already provided Murchison with vital help in identifying and classifying his Silurian fossil collection and in drawing the plates for his 'Silurian System'. He was, Sedgwick told friends, an 'excellent young naturalist' who 'seems quite alive among the old fossil species'.[33] It was probably an indication of the importance Sedgwick attached to the trip that this was the first time he had felt the need to take an expert in palaeontology with him.

I imagine the two men taking the mailcoach along Thomas Telford's 'Irish' trunk road, its top laden with luggage and passengers, through the Midlands and across the River Severn to Llangollen and the sands and limestones of the Dee Valley. From here the road climbs through open pastures and moorland as it skirts the northern fringes of the Arenig hills, before arriving in the disturbed tumble of rocks and upended strata that make up Snowdonia.

Sedgwick's field notes and sketchpads are held in an anonymous red-brick building on the western outskirts of Cambridge. They are infuriatingly difficult to read: Sedgwick wrote in a flowing hand, but his chronic arthritis made holding a pencil difficult; moreover, many of the notes were written as the two men bounced along in their gig. The combination has left a script that is sometimes small and cramped, sometimes large, sloping and virtually unintelligible, with words and lines running into each other. Yet despite these difficulties it's clear the two men spent several weeks in the bleakly beautiful upland moors as the bracken turned to autumn gold, teasing out faint traces of the limestone strata in quarries and river banks across the region, before Sedgwick had to hurry back to Norwich to pick up his cathedral duties.

'We spent many a merry day together,' Sedgwick wrote, 'and every evening, while I slept on the sofa, Salter made the notes, labelled the fossils, made the tea, and woke me up in time to go to bed.'[34] The documents suggest that they located what appeared to be eight or nine faint and intermittent limestone bands running roughly north–south through the hills that stretch from the Berwyns to Snowdonia. The most westerly, which they assumed to be the oldest, outcropped in Cwm Idwal, a hanging valley high on the northern slopes of Snowdon where a glacial lake is surrounded on three sides by steep hills. It was here that Bowman had found the remains of *Asaphus buchii*. Forty miles to the east, on the flanks of the Berwyn Hills near the village of Glyn Ceiriog, a string of spectacular cliffs and limestone quarries running along the valley of the River Ceiriog were logically the youngest. In between were some half-dozen elusive seams that were presumably intermediate in age.

Most required considerable imagination even to see them.[35] I found them impossible to trace, but it was clear that Sedgwick, normally so exacting in his fieldwork, was determined to use whatever evidence he could find to support his progressionist ideas. And yet the results were disappointing. The limestone outcrops yielded some fifteen to twenty new species of trilobite, brachiopod and crinoid, and although many were badly damaged and hard to identify, one thing was abundantly clear: as the two men travelled west to east across the limestone strata, revisiting sites like the quarry at Gelli-grin, there was no perceptible change in the fossil record. By the end of the tour Sedgwick was forced to admit that his initial hope of finding 'certain definite groups indicating a succession in the ascending steps of a vast section ... [had] failed utterly'.[36] The Cambrian System was still without a firm foundation. Meanwhile the Silurian System was looking ever more formidable.[37]

In 1846 the fruits of Murchison's two tours of Russia were published. *The Geology of Russia in Europe and the Ural Mountains* was, as one reviewer put it, another 'monster publication'.[38]

It was 'a difficult book to ignore,' wrote a more recent commentator. 'Tipping the scales at thirteen pounds, filled with fine lithographs of fossils and scenery, it was just the sort of comprehensive monograph that most geologists dreamed of producing.'[39] Praise was forthcoming; one reviewer compared Murchison to Copernicus and suggested that *The Geology of Russia* was 'an achievement of the same order as that which placed the sun in the centre of our system.'[40]

The first volume of two, written primarily by Murchison, described the rocks of western Russia as a vast undisturbed sequence of strata running from the very earliest, primary rocks before life had begun, through the Silurian, Devonian and Carboniferous periods to the Permian and later. They lay mostly in neat, unbroken beds, one on top of the other, apparently the best and most complete record of the history of the Earth ever recorded. What was missing, of course, was the Cambrian, but in the absence of a distinctive fossil signature, Murchison maintained that he had found nothing to support its existence. The detailed evidence to back this up was fleshed out in Volume 2, written in French, largely by Édouard de Verneuil, which included a catalogue of the fossil record underpinning the different strata broken down by family – crinoids, molluscs and the primitive fish fossils of the Devonian.

The Geology of Russia was the definitive account of the Greywacke: the most complete account so far of the Earth's older rocks. In its breadth of coverage it was, in some respects, the book that Sedgwick had set out to write fifteen years earlier, when he had agreed to co-author the second volume of Conybeare and Phillips's *Outlines*, and it gave an air of finality to an 'extended' Silurian System. In doing so, it confined the Cambrian to the margins of geological time.

Tributes to the great work and its author poured in, and

with them 'the honours Murchison had craved since his days in the Peninsular Wars'.[41] 'With the publication of "Russia and the Ural Mountains"', wrote his biographer Archibald Geikie, 'Murchison's position in the very front rank of geologists was universally acknowledged.'[42] The Russians loaded him with tokens of esteem, including a diamond-studded snuffbox and a 4-foot-tall aventurine vase sparkling with particles of gold. The tsar, Nicholas I, also made Murchison a member of the Imperial Academy of Sciences, a post that carried with it the right to be referred to as 'The Right Honourable'. Murchison, for whom such titles mattered a good deal, was delighted and anxiously sought advice on how he might display this honour in his home country. (The solution came later that year when, perhaps prompted by the Russian honour, Queen Victoria knighted Murchison for services to science, allowing him to call himself 'The Right Honourable'.)

Sedgwick wrote to congratulate him. 'Your honours will sit well because they have been hard-earned. They are the fruits of toil and much good service.'[43] Friends began to address Murchison, only half in jest, as 'Lord Greywacke', 'Count Siluria' and even 'Count Silurenski Ouralowski'.[44] One of his greatest admirers was the Lyme Regis fossil-collector Mary Anning, who had come to regard him somewhat extraordinarily as her 'idealised man'[45] – 'certainly the handsomest piece of flesh and blood I ever saw'.[46] Among her possessions was a poem written by a friend and fellow Murchison admirer that ran as follows:

Who first survey'd the Russian states?
And made the great Azoic date?
And worked the Scandinavian slates?
Sir Roderick.

Who calculated nature's shocks?
And proved the low Silurian rocks
Detritus of more ancient blocks?
Sir Roderick.[47]

It says much about Murchison's social status, and perhaps his vanity, that around this time he commissioned Edward Landseer, Queen Victoria's favourite artist, to paint Lady Murchison's dog; the resulting portrait was apparently 'a gem of admirable finish.'[48]

Finally, three years later, at a meeting of the British Association for the Advancement of Science in Birmingham, Murchison received what must have felt like the ultimate tribute. In a special session held in a giant, gas-lit, underground limestone working below Dudley Castle, he addressed an admiring and enthusiastic audience of some 15,000 people on the significance of the Silurian System.[49] 'At a signal,' reported the *Illustrated London News*, 'red and blue fires were lighted at various parts of the caverns, the effect of which was striking and magnificent in the extreme, and drew forth shouts of admiration from the crowds who thronged the caves.'[50]

A flavour of Murchison's new status is apparent in the following account of the occasion:

> In proposing a vote of thanks ... the Bishop of Oxford (Dr Samuel Wilberforce) said that although Caractacus was an old king of part of the Silurian region, yet Sir Roderick Murchison had extended the Silurian domain almost illimitably, and it was only just and proper that there, upon a Silurian rock, he should be acknowledged the modern King of Siluria. The Bishop, then taking a gigantic speaking trumpet, which he had brought with him, called upon all present to repeat after him the words Hail – King – of

> –Siluria! Then, after a pause, the words were repeated a second and third time. The vast assembly thrice responded with stentorian voices and most hearty hurrahs, and ever afterwards Sir Roderick Murchison was proud to be acknowledged 'King of Siluria'.[51]

Charles Darwin would later remark that 'the degree to which [Murchison] valued rank was ludicrous and he displayed this feeling and his vanity with the simplicity of a child'.[52] Typical of this attitude was Murchison's delight in receiving an invitation to brief the King of France, the soon-to-be-deposed Louis-Philippe, on his recent geological travels. He hurried across the Channel 'in full uniform and my Silurian System in hand', recording in rapturous detail his 'most affable and kind' reception at the French court:

> My carriage drew up close to a side-door, which at once opened into a small room in which several clerks were writing, as in a counting house, one of whom told me to sit down. Presently a livery servant appeared, and asked me, 'Est-ce que Monsieur vient voir le Roi?' and then told me that the King would soon see me. After waiting a few minutes, he returned, saying, 'Le Roi vous verra!' and opening a door from this writing-shop, there was the King, who literally seemed to open the doors for me himself. No chamberlain, no officer, not even a sentry between the King and an archway by which all the people passed.[53]

To many in the geological establishment, Murchison's delight in rank and royalty was becoming a source of amusement. 'Roderick Impey Murchison is too omnivorous of fame, grasping at what does not belong to him,' wrote a contemporary.[54] But it had a more serious side-effect. His new status meant that his dogmatic advocacy of a classification of the Earth's strata based entirely on fossils now won almost universal acceptance. So did its associated idea of an 'extended Silurian System'. The publication and enthusiastic reception of *The Geology of Russia in Europe and the Ural Mountains* marked the point when Murchison and Sedgwick's 1834 'gentleman's agreement' on the boundary between the Cambrian and Silurian Systems finally – and very publicly – collapsed. For ten years the boundary between the two men's 'territory' had been more or less successfully maintained. Now the fact that the Upper Cambrian Bala Limestone and Berwyn Slates, and the Lower Silurian Caradoc and Llandovery rocks appeared to share an almost identical lithology and fossil fingerprint meant that, for most people, the idea of rolling them together into an extended Silurian System was only rational. 'Mineralogically there was really no true boundary line,' Murchison wrote at the time, 'and zoologically it had been agreed even by Sedgwick himself that no distinct assemblage of fossils had been ascertained to belong to the Cambrian series.'[55]

John Phillips, the highly regarded professor of geology at King's College London, who had long opposed Murchison's simple stratigraphy, urged caution.[56] 'I wish to put again and to keep for some time in doubt the exact lines of demarcation which best suit the soft shades of mother nature,' he wrote, arguing that any stratigraphic system that truly reflected the real world needed to be subtler and more sophisticated:

> Murchison's Silurian System is a noble assemblage and he is entitled to every applause that he has got and more but I hope you will have the magnanimity to permit the whole basis and superstructure of the classification

> which he has favoured, to undergo a new and searching analysis for the discovery of perhaps new and important relations.[57]

Phillips was referring once again to the possibility that two geological periods might contain the same fossil record. But few people were prepared to listen. It looked like the end of Sedgwick's Cambrian System.

Soon afterwards Murchison's extended Silurian System received what amounted to official approval. Henry De la Beche's Geological Survey had by now grown into a team of seven or eight young geologists charged with mapping the geology of Britain. In the early 1840s the team, dressed in the obligatory military-style uniform of blue serge with brass buttons and a top hat, had moved into south Wales and over the following years had gradually worked their way north. Now, in 1851, the team published what many regarded as the definitive map of central and northern Wales. It was instantly clear that it had been heavily influenced by Murchison's concept of an extended Silurian: the map rolled the Upper Cambrian and Lower Silurian periods together as a single series of undulating bands of shelly limestone, confirming their belief that the terms Upper Cambrian and Lower Silurian were two names for the same series of rocks.[58] It labelled them 'Silurian'.

Sedgwick's Cambrian System was left with only the very oldest, fossil-free Snowdonia Slates: the base of the fossil-bearing Lower Silurian rocks. 'Your Silurian System must have a jolly extension at our hands over the rocks of this land,' wrote De la Beche to Murchison, who after the fierce fighting over the Devonian rocks had no particular reason to look favourably on him.[59] It was the official endorsement of what many geologists had already accepted.[60] To Sedgwick, it was an enormous

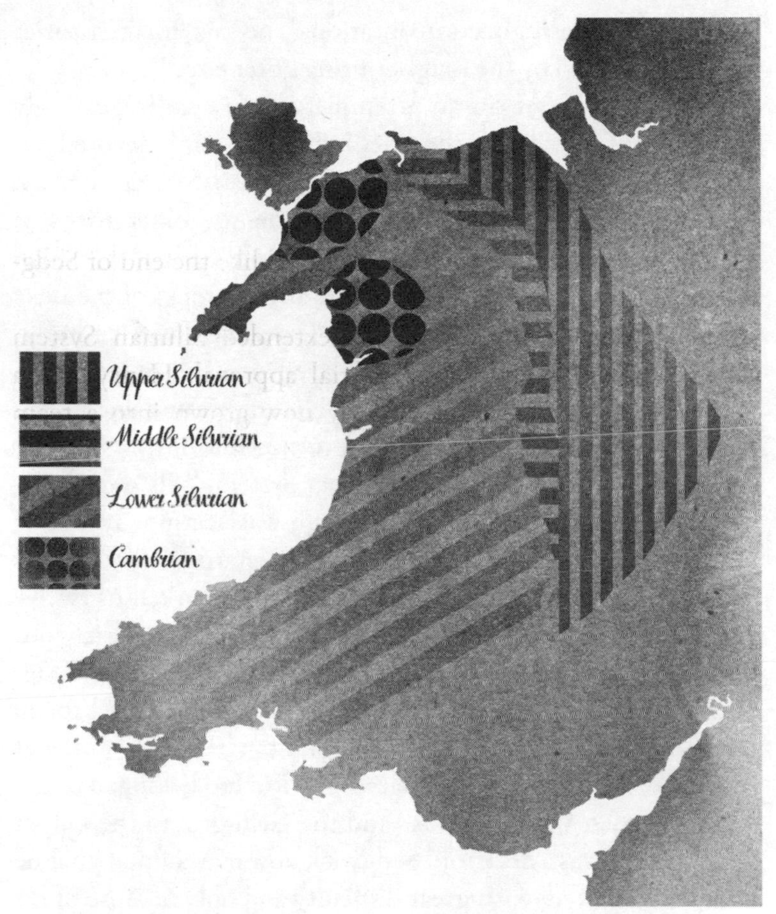

7.2 A simplified map of the boundary between the Cambrian and Silurian, according to the work of the Geological Survey in 1852. It drew heavily on Murchison's classification of the rocks.

blow and he was outraged at what he saw as a personal betrayal: his Cambrian System had been confined to the peculiar slatey rocks of a tiny corner of north Wales, rather than a major period in the unfolding of the Earth's history. In a language that was becoming increasingly confrontational, he saw 'his' territories as being 'annexed by the neighbouring potentate'.[61]

In February 1852, in an attempt to defuse an increasingly acrimonious situation, the Geological Society devoted an entire meeting to the issue of the Cambrian/Silurian border. It included almost all the specialists on the older rocks in Britain.[62] The centrepiece was a paper from Sedgwick, which ostensibly drew parallels between the ancient rocks of the Lake District and those in north Wales.[63] In reality it was 'an indignant protest' against what he perceived as his ill treatment at the hands of the Geological Survey. His anger burst uncontrollably to the surface,[64] and the paper turned into a frontal attack on Murchison and his entire Silurian System. 'His nomenclature was premature and his base line was sectionally wrong and so far from leading to discovery, it retarded the progress of Palaeozoic geology for, I believe, not less than ten or twelve years,' he told the meeting ('Palaeozoic' being the generic word for the earliest geological periods).[65] The great mistake, Sedgwick went on, had come in 1834, when Murchison had broken the correlation between the Meifod and the Bala Limestones at places like Gelli-grin, and declared that Meifod belonged to the Lower Silurian Caradoc rocks and the Gelli-grin to the Upper Cambrian. It was a decision, Sedgwick now maintained, that he had only accepted 'with great difficulty and only because of his implicit faith in the perfection of Murchison's workmanship'.[66]

The meeting was 'very warm', wrote one observer and lasted beyond midnight.[67] 'The excitement was great and the discussion animated. It was long since such a bolt had fallen out of the blue,' recorded another.[68] Sedgwick 'gave forcible expression to his dissent, using language with respect to his old comrade, which ... [was] felt by the friends of both antagonists to be too personal'.[69] It raised a storm of protest. 'Good scrimmage between S[edgwick] and M[urchison] on the L[ower] S[ilurian] and C[ambrian] question,' wrote Andrew Ramsay, a leading member of the Geological Survey, in his diary. 'It was not an enlivening spectacle. Sedgwick used very hard words.'

Murchison, reluctant to break with his friend and colleague of twenty-five years, tried to make light of the affair, suggesting that 'the only rationale is that he is in love & has lost his head like many great men before him ... so says rumour!'[70] Others were less charitable, hinting that Sedgwick, now in his late sixties, was growing senile. 'I had a very long talk with [Sedgwick] & a very friendly one at the Athenaeum two days ago,' wrote Edward Forbes – another young recruit to the Geological Survey, and its leading palaeontologist – to a colleague, 'but found him so vague and dreamy & so ignorant of the real natural history question that I did not press any discussion.'[71]

Two days later Murchison wrote to Sedgwick denying again that he had acted unfairly. 'This is the only point which roused my feelings ... and made me speak more vehemently than I intended,' he wrote. 'It was not me, I repeat, who made *Cambrian into Lower Silurian*, but the government surveyors and palaeontologists' (emphasis in original). And he concluded, 'I have been grievously pained to be set in antagonism to you but I can solemnly assure you that I know no possible way by which my present position could be altered without stultifying my original views of the Silurian System as a whole.'[72]

The row overflowed into the pages of the *Literary Gazette*, a literary and scientific weekly. This 'led to an exchange of published letters repeating all the old arguments ... but now in the full glare of the public gaze'.[73] It also appeared in the *Edinburgh Philosophical Journal*. Over the following months the two men continued to correspond in relatively amicable terms. 'I can safely aver that I did nothing whatever to induce the Government surveyors to adopt the line they have, and I never went into your region until they had quite settled all their nomenclature,' Murchison wrote to Sedgwick in November 1852.

> It has a very bad effect upon the progress of our science to see Sedgwick and Murchison trotted out as controversialists. All our oldest and best friends regret it sincerely, and the more so as there is really nothing in the philosophy of the case on which we differ. We agree in the grand doctrine of a progression of creation [as opposed to Darwin's evolution] and we both start from the same point, now that the data are as fixed in the British Isles as they are in other countries.[74]

But there was no hiding the growing animosity between them. It was at this moment, wrote an observer, that 'the seeds began to germinate' of an estrangement that would begin to eat away at 'the ancient brotherly friendship between Sedgwick and Murchison'.[75]

For all his anger, the situation was largely of Sedgwick's own making. While Murchison had produced one major tome after another, Sedgwick had made almost no progress on his own book on the Cambrian. Colleagues urged him to hurry up. 'I'm anxious you should put on deliberate record some abstract of your labours,' Beete Jukes wrote to him, as the government's team of surveyors worked their way north through Wales. 'Your contemporaries know how much you have done as a pioneer in opening up this department of science, but the rising generation of geologists do not.' At the very least, Jukes urged Sedgwick, 'you should publish some condensed well considered work that can be appealed to by *name* hereafter' (emphasis in original).[76] Murchison also continued to press him. 'I have over and over urged you to bring out your fossils and complete the subject you had undertaken. It was no fault of mine that you did not do this.'[77]

But Sedgwick had made almost no progress on his Cambrian book for at least six years: 'I have neither time nor strength for the big Palaeozoic work I have long contemplated,' he wrote back to Beete Jukes. 'I'm buried in the enormous mass of my materials.'[78] 'On the principal geological issues to which he had devoted his life, Sedgwick was becoming an isolated dissenter,' writes James Secord, 'beleaguered by … the widespread desire for an unchanging classification useful in international correlation.'[79]

Two decades earlier, at the start of their collaboration, Sedgwick had been the senior partner; now the roles had been reversed. Nothing showed this more clearly than a poll conducted at around this time by the Institute of France, the leading learned society in the country, to determine who should become the British corresponding member. Sedgwick received three votes while Murchison was elected with twenty-seven – more than all the other candidates put together.[80] 'By the early 1850s,' writes James Secord, 'Sedgwick's program of research was on the verge of failure, impeded by his inability to undertake sustained literary composition.'[81]

What lay behind this failure to put pen to paper? Did Sedgwick worry perhaps that the ideas generated by the new science, shedding light on the age and history of the Earth – not to mention the origins of life – might clash with his religious faith? It seems possible. He was certainly, like many of his generation, unable to accept Darwin's theory of evolution. But by the 1840s and early 1850s most clerical geologists had come to adopt a somewhat double-minded approach to the problems that geology had thrown up for religion, accepting that both, in their different ways, could be 'true'. It helped that most of them were very consciously empiricists, primarily interested in naming, classifying and cataloguing rocks, bringing order to the apparent chaos, rather than concerning themselves with theories about how and why the rocks had come into existence. George Greenough spoke for many when he urged 'a strictly empirical approach' to the subject, and insisted that the basic concepts of the new science had 'to be purged of all unacknowledged theoretical overtones' before any advances could be made.[82]

Alternatively, perhaps Sedgwick was simply unable to curb the noise in his head generated by the pressures and events of everyday life, and to concentrate on a single idea for long enough to commit it to paper.

Chapter Eight

FROM COLLABORATION TO RIVALRY, 1852–5

When the minutes of Sedgwick's stinging attack on Murchison were published in the Geological Society's quarterly journal in May 1852, they sent shockwaves through the readership. 'There are statements and expressions which Sedgwick ought not to have used ... and which ought still less to have been printed in our journal,' stated one appalled recipient.[1] When the 'pungent language, stripped of all the humour and bonhomie of oral delivery, came to be calmly read in print, there was a very general expression of sympathy with Murchison,' wrote another.[2] The administrative fathers of the Geological Society panicked and, a week later, 'in their anxiety for peace', as one observer put it, took the unprecedented step of attempting to remove 'the unclean thing' from their journal by cutting the last fifteen pages.[3]

It was a decision unique in the annals of Victorian science; and 500 copies of the journal had already been circulated. Sedgwick was outraged: it constituted 'a personal stigma unexampled in the history of any philosophical society in London,' he fulminated and left 'an indelible sense of wrong' in his mind.[4] His supporters lined up behind him to condemn the move. Relations between Sedgwick and the Geological Society became increasingly strained.

Behind the scenes, however, there had been a new

development at the Woodwardian Museum. For several years Sedgwick had been employing an energetic young Irishman called Frederick McCoy to help him organise the fossil collection. McCoy was bright and precocious: at the age of eighteen he had published a catalogue of the entire fossil collection held in the Rotunda, a public exhibition space in Dublin. Sedgwick regarded him as 'an incomparable and most philosophical palaeontologist' and 'one of the steadiest and quickest workmen that ever undertook the arrangement of a museum'.[5]

McCoy had begun by arranging the Cambridge museum's collection of Cambrian, Silurian and Devonian fossils in the traditional manner, according to their different zoological groupings: the dozen or so trilobite species in one area, the equally numerous brachiopods in another, the crinoids, bivalves and coral species in yet others. Then, as an experiment, he rearranged them according to where in the rock strata they had been found: in particular whether they had been found in what Sedgwick was still calling his Bala beds, at places like Gelli-grin in the Berwyn Hills, or in Murchison's Caradoc Sandstones at sites like the quarry at Meifod.

The result was a total surprise. When McCoy looked closely at the Caradoc fossils he discovered they fell into two distinct groups. One group shared fossils with Sedgwick's Bala rocks, while the other contained numerous fossils that were also found in Murchison's Upper Silurian Wenlock formation.[6] There appeared to be no overlap between the two groups: not a single species bridged this divide. If McCoy was right, it split Murchison's Caradoc system into two groups with two entirely different fossil fingerprints, which in turn – according to all the rules of fossil-based stratigraphy – implied two distinct periods in the Earth's history: what McCoy now began to call an Upper and Lower Caradoc. This, in effect, demolished the concept of the Caradoc as a coherent geological period. The upper part seemed to have more in common with the rocks above, and the lower part with those below, while between them there appeared to be a complete change in the rock record.

Sedgwick's biographers, Clark and Hughes, offer no insight into Sedgwick's reactions to this remarkable find, but it's safe to assume that he was intrigued. Although he had spent years arguing against too great a dependence on fossils to classify rocks, this looked like a major new insight. During the early summer of 1852, while outwardly still fuming over his treatment at the hands of the Geological Society, Sedgwick laid plans to return to Wales in search of further evidence.

But that summer it was hard to leave Cambridge. For some time Sedgwick had been serving on a government-appointed inquiry into the funding and administration of Oxford and Cambridge Universities.[7] There were concerns about the standard, relevance and cost of degrees and the quality of the teaching. In addition there was a growing unease about the quite extraordinary power the two universities exercised over their surrounding non-student population, including the right to arrest and punish prostitutes, license alehouses, supervise the weights and measures in markets and license theatres and entertainments.[8] Reform was overdue and Sedgwick, as a senior academic and proctor, was deeply involved, writing long sections of the report on Cambridge. By the summer of 1852 the inquiry had reached its final stages and Sedgwick, never the most diligent writer, was forced to drop everything to complete his contribution.

It must have been a galling experience. The weather that summer was exceptionally hot – a good time for fieldwork in Wales – and he was sitting on potentially explosive new findings. 'I am smoking at every pore,' he wrote in frustration to his niece

Fanny Hicks. 'The perspiration hisses from me with a noise like a steam-whistle; I am in a state of fusion and confusion – ever moping and mopping, a miserable walking *machine a vapour*. Sometime I think my poor bedmaker will find me running all over the floor and have to pick up my organic remains in slop-pails' (emphasis in original).[9]

July and August went by, and so did early September. The summer window when the weather in Wales was reliable enough for fieldwork was vanishing rapidly. Moreover, Sedgwick had to be back in Cambridge by the beginning of October to supervise the annual Fellowship exams. This left him around ten days in the field.

The original idea had been to revisit some of the classic outcrops of Caradoc Sandstone in north Wales and Shropshire, but now he decided on a new and less ambitious route. For some time there had been reports of a puzzling band of Caradoc Sandstone on the slopes of May Hill, a prominent landmark at the southern end of the Malvern Hills in Gloucestershire, where a survey by his old ally, the palaeontologist John Phillips, had uncovered a band of rock that seemed to occupy an ambiguous position in the stratigraphic table. Its structural position suggested that it was part of Murchison's Upper Silurian Wenlock formation, but on the basis of its fossil fingerprint it had been assigned to his Lower Silurian Caradoc rocks. Sedgwick was intrigued: could it be that two different 'Caradoc-type' rocks had been confused and rolled into one – a classic example of McCoy's Upper and Lower Caradoc?

May Hill rises proudly out of the surrounding farmland to a height of nearly 1,000 feet, part of a range of hills that separate the valleys of the rivers Severn and Wye. For hundreds of years it had been a beacon for ships navigating the ever-shifting sands of the lower reaches of the Severn on their way to inland ports at Gloucester, Tewkesbury and Worcester: a comforting vision for returning sailors of a familiar and unchanging landscape. As the local poet John Masefield wrote in his 1911 poem 'The Everlasting Mercy':[10]

> I've marked the May Hill ploughman stay
> There on his hill, day after day
> Driving his team against the sky,
> While men and women live and die.

Nowadays a small plantation of ancient pines on the summit offers a sense of the same deep continuity. On clear days the views from the top of the hill are magnificent: to the south the silver-grey meanders of the lower Severn and the distant Cotswold Hills; to the west a series of parallel ridges that stretch to the distant horizon; and to the north the rough grasslands of the Woolhope Dome and the Malvern Hills.

But Sedgwick and McCoy would have seen none of this. After a blistering hot July and August, the weather broke and low clouds and rain obscured the landscape. Sedgwick's field notes for the trip have disappeared, so it's unclear where exactly the two men went, but on the basis of the area's geology it's likely they found John Phillips's ambiguous 'Caradoc' rocks on the southern flanks of the hill.[11]

It was exactly as McCoy had predicted. Running through what had been classified as a single bed of 'Caradoc' sandstone the two geologists found a break or unconformity in the beds of rock. Two strata, superficially similar and lying next to each other, had been rolled into one and dubbed Caradoc. This error had obscured a change in the Earth's environment so great that one set of life forms had vanished completely and an entirely new one had appeared: the sort of change usually associated with the transition from one geological period to another.

There had also been reports of similarly confusing Caradoc rocks further north in the Berwyn Hills. However, that September, after tramping up May Hill in the wet, Sedgwick took to his bed with a streaming cold and remained there for most of the time remaining for the trip.[12] Ten days of fieldwork had shrunk to two. McCoy passed the time visiting nearby outcrops of Caradoc Sandstone, but the results were inconclusive.

Sedgwick returned to Cambridge in the autumn of 1852 convinced, however, that he had enough field evidence to support McCoy's claims of a stratigraphic break at the heart of the Caradoc rocks.

Two months later, at the November meeting of the Geological Society in London, Sedgwick and McCoy went public. In a paper provocatively entitled 'On the Separation of the so-called Caradoc Sandstone', Sedgwick split the Caradoc into two periods, separated by an unconformity that, he argued, was much greater than that between the Silurian and the Devonian, where the two systems shared several life forms. He assigned everything above the unconformity to a reduced Silurian System; everything below it he claimed for his Upper Cambrian Bala/Llandeilo beds, creating in effect an 'expanded' Cambrian System. It meant that for the first time he could finally lay claim to a unique set of fossils for his Upper Cambrian rocks, including those old friends from the Gelli-grin, the brachiopod *Orthis fabellulum* and the trilobite *Trinucleus caractaci*. The Cambrian was no longer confined to the oldest, fossil-free rocks before life had begun.

In the small world of British geology this 'was nothing short of sensational', writes the historian James Secord.[13] It was not only an attack on Murchison and his view of the Silurian System; it also challenged the Geological Survey, which had so ostentatiously adopted Murchison's extended Silurian System and used it to draw what was generally regarded as the definitive geological map of Wales. If Sedgwick and McCoy were right, then large areas of rock currently assigned to the Caradoc would have to be reassigned and the map redrawn, just twelve months after it had been completed.

Most members of the Geological Survey, and much of the geological establishment, lined up with Murchison at the meeting, in greeting Sedgwick's new paper with considerable scepticism. Many doubted the accuracy of McCoy's palaeontological work. 'The impression left not only on my mind,' reported Edward Forbes, a young naturalist and recent recruit to the Geological Survey:

> was that McCoy ... had cooked the fossil evidence to please Sedgwick & had misled him. Salter [for the Geological Survey] spoke very well and from good data. His speech was very convincing and went to show that the Survey was justified in ... [its existing mapping]. There was the usual spar between Murchy & Sedgwick about Cambrian. The discussion was lively & good.[14]

Surprisingly John Phillips, who had first reported the ambiguous strata at May Hill, spoke against Sedgwick, accusing him of being an 'ultra palaeontologist' and relying too heavily on the local fossil evidence, a charge that sat oddly with Sedgwick's known doubts about fossils.[15] The debate is said to have continued until nearly midnight.[16] Even afterwards the Geological Society delayed publication of Sedgwick's paper until it had been vetted and double-vetted 'to exclude all controversial matter'.[17]

Yet outside Britain, Sedgwick's new stratigraphic table made a lot of sense. The American geologist Henry Darwin Rogers,

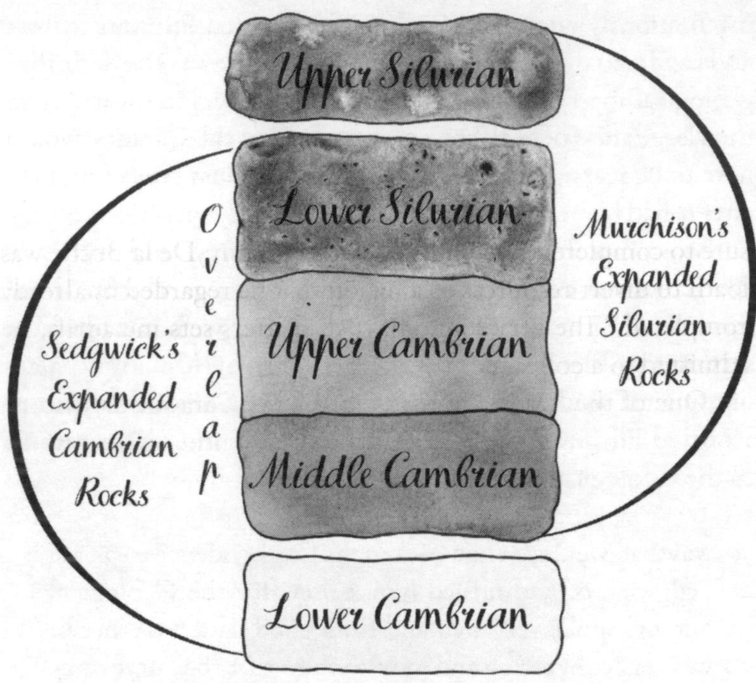

8.1 Sedgwick's 1852 expanded Cambrian System.

one of Sedgwick's first and most important US allies, wrote that the break in the Caradoc was 'infinitely better in harmony with our American Paleozoic Geology' than Murchison's, and informed Sedgwick that his rearrangement of the Caradoc rocks would be 'more acceptable and intelligible to the American geologists than any other scheme of arrangement of the British rocks which has yet been published'.[18] It also received wide support from a number of geologists across Britain who had noticed similar anomalies.[19] As a result, within months of the meeting there was a growing sense that this was an important new development and that Sedgwick's claims for an expanded Cambrian System warranted further investigation.

Reluctantly and with a heavy heart, the director of the Geological Survey – the now far grander Sir Henry De la Beche[20] – gave the go-ahead for members of his survey team to return to the Welsh borders to review some of the classic outcrops of Caradoc rock. It can't have been an easy decision. The survey was surviving on a tight budget and was under considerable pressure to complete its geological map of Britain; De la Beche was loath to divert resources to a region that he regarded as already completed. 'The very sound of such matters sets me adrift,' he admitted to a colleague.[21]

One of the better-known exposures of Caradoc Sandstone was in a small cliff on the River Onny in southern Shropshire, near the tiny farming settlement of Cheney Longville. Here, in a countryside of small woods and gently sloping fields of cattle, the fast-flowing river had exposed bands of pale-yellow sandstone and shale. Murchison had always considered it a classic example of Lower Silurian Caradoc Sandstone ascending smoothly, without a break, into Upper Silurian Wenlock beds. Sedgwick and McCoy had paid the site a fleeting visit the year before on their way home from the Malverns, but had been unable to see the river cliff because heavy rain had raised the water level. Sedgwick predicted an unconformity, but left without any evidence to support it. Now, in the summer of 1853, De la Beche sent two of his most experienced geologists, William Aveline and John Salter, to re-examine the cliff.

Aveline, a 'tall, dark, silent, big booted man who strode with gigantic steps over the hills', had been part of the team that had already surveyed much of Wales.[22] John Salter had worked closely with both Sedgwick and Murchison and was a highly experienced palaeontologist. They must have been enormously competent at reading the rocks. I visited the quiet valley one late-summer afternoon, following a grassy track through ivy-draped woods of oak and sycamore to the river bank. Small trees and shrubs had taken root in the sandstone, obscuring much of the cliff. But even knowing what I was looking for, I doubt I would have seen anything unusual. The strata dip, gently and apparently uniformly, to the south-east, a small change in colour halfway up marking the boundary between what had long been regarded as the Lower Silurian Caradoc and the Upper Silurian Wenlock rocks. But Aveline and Salter saw something that everybody else had missed. Instead of dipping uniformly, there was a slight variation in the dip between the upper and lower strata, suggesting there had been some kind of break in the deposition record and that the bottom strata had been subject to a period of erosion before the top strata had been laid down.

'It is difficult to examine as, except in droughts, one has to stand up to the knees in water,' the two geologists reported to a subsequent meeting of the Geological Society. All the same, they were able to establish that there was indeed 'a great unconformity between the Upper Silurian series and the subjacent so-called Caradoc Sandstone'. There is '*no trace of passage* nor any similarity in fossil content,' they wrote, 'and it is evident that they can no longer be classified under the same [Silurian] name' (emphasis in original).[23]

It was exactly what Sedgwick had anticipated. McCoy couldn't help crowing, 'Salter and Aveline's paper is a most complete admission of our (much contested) views of the Caradoc sandstone ... it is great fun to see the Survey marking out the May Hill plane of separation, after all their stout contradiction and opposition.'[24]

The dam of opposition had burst. Over the following months small teams from the Geological Survey visited dozens of exposures of Caradoc rock at places like the slate-mining village of Glyn Ceiriog on the north-eastern slopes of the

Berwyn Hills, and the hamlet of Mathrafal on the south flanks of the hills where the River Banwy flows over a shallow shelf of rock, exposing low banks of sandy conglomerate. Again, to the untrained eye, the unconformity can be hard to spot, but time and again Geological Survey members found the same, previously unnoticed gap in the rock record that split the Caradoc into two.[25] 'The failure to observe the unconformities like those at Glyn Ceiriog, Mathrafal and Onny is easy to understand,' wrote a later commentator, 'for in many cases these exposures were obscured and the exact points of contact difficult to see on rapid examination.'[26]

The consequences, however, were enormous. From north to south Wales and through the Malvern Hills, adjustments had to be made to the boundary between the Sedgwick's Upper Cambrian and Murchison's Lower Silurian. For the first time in its twenty-year life the Silurian System was coming under serious threat. Geologists – nearly all of them supporters of Murchison's vision – were being forced to admit that they might have got it wrong. With a distinct fossil fingerprint of its own, the Cambrian Period appeared to have finally achieved the same standing as the Silurian, Devonian, Carboniferous and Permian periods, while the Silurian had been chopped in half.

By the late summer of 1853 two competing and very different maps of the geology of Wales were in circulation. One, produced by Sedgwick, showed essentially an expanded Cambrian System and a much-diminished Silurian. The other, drawn up by Murchison, showed almost the exact opposite: an expanded Silurian, with the Cambrian confined to a local slate-heavy subdivision of the Lower Silurian, without a colour of its own.

The news hit Murchison at a bad time. For several years he had been preparing a condensed and more popular account of the early rocks of the entire world. *Siluria*, as it was called, was nearly finished and the pages on the Caradoc were about to go to the printers. The new evidence, backed by the Geological Survey, appeared overwhelming, but Murchison had come to regard the rocks in Wales as notoriously confusing and difficult to read, and he was unwilling to be dragged back into what he regarded as an old, irrelevant debate on the basis of what might be regarded as a quirk in the Welsh strata. He was now working on a global scale, convinced that the Welsh findings were merely a local phenomenon.

It was a defence that he would keep making for the rest of his life.

Throughout the remainder of 1853 and the following year Sedgwick and Murchison clashed time and again. It began relatively amicably, but soon degenerated: 'In looking over our old publications and my old memoranda-books,' Murchison wrote to his former partner in the autumn of 1853, 'we seem to be such complete Siamese twins that it does my heart good to turn to them and pass away from all the irritamenta about a nomenclature which has led too many persons to think that we were estranged. I will never go on wrangling. What I have done, and said, and published, has never impugned the accuracy of your labours in the field, and I only regret that some expressions that have fallen from you about my mistakes and errors should have appeared harsh to others.'[27]

But Sedgwick's reply suggests he was still brooding over Murchison's attempts to annex 'his' Cambrian System for Siluria:

> I know that I am a great procrastinator [for not completing my book], partly from temperament, partly from multitudinous engagements that pull hard at me, and chiefly from a condition of health which for months and

> months together makes writing and sedentary work very very irksome, and almost impossible. Still, though a man is behind time with his rent, he rather grumbles when he finds on coming back to his premises that a neighbour has turned out his furniture, taken possession, and locked the door upon him. This is exactly what you did, and so unexpectedly, and without notice.[28]

By now, in any case, it was clear they were fighting not only over territory, but also over different versions of what they considered the truth. Sedgwick was convinced that the break at the heart of the Caradoc strata in places like the Onny Valley and Glyn Ceiriog represented a universal finding. Murchison was equally convinced it was merely a local anomaly, with no bearing on the Caradoc elsewhere.

Young geologists, full of pride in their new science and the camaraderie that came from weeks spent together in the hills, feared that the dispute would degenerate into a bitter slanging match. 'Allow me to hope gentlemen that along with the scientific and economical benefits which our science has conferred upon the world, it may still remain the boast of Geology that she has contributed her share to the moral benefit also,' the surveyor Beete Jukes told a meeting of geologists in Dublin, and 'that philosophers can differ in opinion without loss of temper and without loss of respect for their opponents'.[29]

It was a forlorn wish. Six months later, in the spring of 1854, Sedgwick finally and irrevocably fell out with the leadership of the Geological Society. He had recently completed an updated paper on the May Hill sandstones, which reaffirmed his view of an expanded Cambrian System.[30] The Society's Council, still undecided on the issue, referred it to a third-party referee who suggested amendments. Sedgwick, convinced that Murchison was pulling strings behind the scenes, was furious and sent it instead to the old established scientific journal, the *Philosophical Magazine*. The Council of the Geological Society was, in turn, equally outraged: the publication elsewhere of a paper already submitted to the Society violated one of the organisation's cardinal rules.[31] Sedgwick was publicly censored, and responded in turn by threatening to leave the organisation. The Society's president, William Hamilton, urged him to think again: 'I read with the greatest regret any allusion to the possibility of you withdrawing from the Geological Society. To me personally it would be an especial cause of sorrow if it should take place during my presidency.'[32] But it was too late. Sedgwick, fuming with resentment, never again attended a Society meeting.[33]

Later that year the squabble reached a new low in Liverpool, at the 1854 meeting of the British Association. Both Murchison and Sedgwick produced papers supporting their different positions: Sedgwick's an update of his May Hill paper for the Geological Society. The audience expected a fierce debate, and got it. 'The opposing forces ranged themselves in order of battle and a war of words commenced which must have lasted for several hours,' ran one account.[34] It was, wrote James Secord, 'perhaps the most spectacular verbal pyrotechnics ever seen at an Association meeting'.[35] Sedgwick complained that 'his principality had been invaded without a declaration of war'; and that the 'Silurian colours had swept over it', without him being warned.[36] Many in the audience were said to be shocked by the vehemence of his language. 'It is no wonder that you were pained by the tone and matter of the discussion of the Palaeozoic in Liverpool,' John Phillips wrote to Murchison afterwards. 'Who was not?'[37] 'The language he has allowed himself to use is wholly unjustifiable,' wrote another. 'It is undignified and must be universally condemned.'[38] The Geological Society's Edward

Forbes, who had chaired the session, worried that the publicity given to the row 'would lead the uninformed to think that "all geology had come to a hitch" thereby opening the door to Scriptural geologists who would use the dispute to cast doubt on the whole geological enterprise'.[39]

Constructive dialogue had become impossible. The Geological Society reluctantly called a halt, announcing a ban on the publication of any new papers on either side of the debate. By the winter of 1854 Sedgwick and Murchison had stopped talking to each other. The two men had looked at Wales and come to radically different conclusions. The 1834 boundary between the Cambrian and Silurian was long dead, but nobody could agree about what to replace it with. A thirty-year working friendship had degenerated into bitterness and acrimony.

Neither man was innocent. Murchison, ambitious, self-seeking and often self-important, was not above playing fast and loose with the truth and had displayed a quite extraordinary insensitivity to the feelings of those he regarded as standing in his way.[40] It's almost inconceivable that during the late 1830s and throughout the 1840s he failed to realise that Sedgwick might take offence at his attempts to renege on the 1834 'consensual boundary'. Never the less, he had unilaterally redrawn the border between the Cambrian and Silurian systems, chopping the former in half and consigning it to the very bottom of the stratigraphic table before life on Earth had appeared.

And yet, in Sedgwick, Murchison must have found a profoundly frustrating collaborator: a man whose repeated failure to meet his writing commitments would have driven even a more patient person to despair. Time and again Sedgwick made promises he failed to keep, missed deadlines he had guaranteed that he would meet, and had to be cajoled into committing his thoughts and fieldwork to paper.[41] 'I did put north Wales into good approximate order and ought long since to have published the details,' he confided to a friend some years later, 'but rheumatism, gout etc. etc.... [got to me] so that I have done nothing hardly in the writing way for some years.'[42] Moreover, as Murchison developed in stature, Sedgwick appears to have grown increasingly resentful and quick to take offence, finally retreating into self-imposed exile in Cambridge, convinced that the world was against him and angrily complaining that his Cambrian System was being 'suppressed'.[43] 'I fought my way single handed to a knowledge of the Cambrian Series and I have defended my classification single handed and without fear of consequences,' he wrote self-pityingly, and not entirely accurately, in 1855.[44]

To close friends and relatives, Sedgwick was still capable of great warmth and friendship, but he was growing tired, ill-tempered and subject to increasing bouts of melancholia and loneliness.[45] It is perhaps telling that during a trip to Scotland at around this time he visited Robert Burns's family home in Ayrshire and read with 'huge delight' Burns's classic poem 'Despondency: An Ode':

> Oppressed with grief, oppressed with care
> A burden more than I can bear,
> I sit me down and sigh:
> Oh life! thou art a galling load,
> Along a rough, a weary road,
> To wretches such as I![46]

It was a subject close to his heart. Trips to the hills, once a source of pleasure and escape, had become difficult and solitary; the romance of the rocks had worn thin. 'Miserable damp weather has made me rheumatic and low spirited,' he wrote in a sad, reflective letter to his niece, Fanny Hicks, during one such trip. 'I have during the summer been wandering through lovely country but I am almost in solitude. I had no one to talk to; and I could not multiply my pleasures, or add to their greatness by seeing them reflected from a friend's eye or echoed back by a friend's tongue … I am old and cold hearted and torpid and rickety and there is no help for it.'[47]

He continued to be morbidly preoccupied with his health. 'For 29 days I didn't leave my rooms and I spent my days in drinking slops and soporifics enough to dissolve a block of granite,' he told a friend during the winter of 1853–4. 'Mustard footbaths and mustard chest cataplasms were all in vain. So the doctor dabbed my throat and chest with a liniment which gave me a kind of horrible red mange, and made me unfit for a civilised piggery.'[48] He also talked of 'strange clouds of oblivion which too often trouble an old man's memory'.[49] But above all else his mood was coloured by his failure to fulfil his original promise and write his own version of Volume 2 of Conybeare and Phillips's *Outlines*. After thirty years of hard and often pioneering work he had nothing to compare with Murchison's *The Silurian System* or *The Geology of Russia in Europe and the Ural Mountains*, and it can't have helped that during the summer of 1854 Murchison's *Siluria* was published, presenting his global view of the Earth's early rocks.

It met with a mixed reception. Though large and well illustrated 'with abundant woodcuts and plates of fossils',[50] it suffered from Murchison's tendency to dry and excessive detail. 'Sir Roderick is too intent on the matter of what he has to say to take much pains with mere manner [which] results in a dry and rather uninteresting style, rarely enlivened by any graphic or forcible word-painting,' wrote one reviewer,[51] while another suggested that it 'lacked the freshness and originality' of earlier works.[52]

For many readers, however, it also represented the final triumph of Murchison's extended Silurian System and the years of painstaking fieldwork that had gone into it. 'The energy of fifty hammers, guided by as many good hands, all striking together, could not exceed that of the man who founded the Silurian System,' wrote a reviewer in the *Literary Gazette*. It is 'one of the greatest achievements that have been made in our time towards the consolidation of geological science'.[53] To Sedgwick, it must have seemed like yet another reproach for his own failings.

Yet to a new generation of geologists the battle between Sedgwick and Murchison was coming to seem like an obstruction to progress. I am 'bored and bored again with a set of arguments 50 times repeated,' Andrew Ramsay, a long-standing member of the Geological Survey, wrote to a friend.[54] Many younger geologists were beginning to wonder whether, in light of the discovery of a break in the Caradoc strata, the rocks had a more subtle and complex story to tell, which was being obscured by the noise and clamour of the rivalry. Was there something else going on in the twisted and contorted rocks along the Cambrian/Silurian border? A new clue – possibly a misunderstood fossil – that might explain the apparent impasse? The initiative was about to pass to a new generation of geologists who were less invested in the old arguments.

Chapter Nine

THE HIGHLAND CONTROVERSY: SILURIA IN SCOTLAND, 1855–73

In early August 1855, with glorious summer skies above him, Murchison travelled north. He probably took the recently opened steam-train service between London and Glasgow and then hired a coach and horses to follow a track beside Thomas Telford's Caledonian Canal, through a wild landscape of mountains and lochs, to the north-eastern city of Inverness. From there he continued north to the small market town of Dingwall, skirting the tranquil shores of Beauly Firth and his old family home at Tarradale House, sitting 'quiet and peaceful, nestling down among its old trees'.[1] Sixty years of often-brutal land 'Clearances' had dramatically altered the landscape. Thousands of tenant farmers had been swept from the surrounding hills in the name of agricultural progress, and small mixed farms had been replaced by large, open fields of grazing sheep.[2] 'Cottagers, who had formed so conspicuous a feature of the district in early days, were all gone,' Murchison observed with a mixture of nostalgia and pride. 'The little huts and rude stone walls had been cleared off to make way for the large well-tilled fields.'[3]

Murchison was accompanied by an eager young colleague called James Nicol, a former assistant secretary to the Geological Society in London, who, with Murchison's help, had recently been promoted to the chair of natural history at Aberdeen

University. Nicol was an enthusiastic proponent of Murchison's ideas. Some years earlier he had published a series of papers on the poorly understood rocks of the Scottish borders, the so-called Southern Uplands, which suggested that they were largely undisturbed beds of Lower Silurian rock. The fossil record was poor and consisted largely of minute scratchings in the rock, thought to be the remains of small plankton-like creatures known as graptolites – tiny flakes the size and thickness of a fingernail, some people doubted were fossils at all. But Nicol had been able to find just enough evidence, mostly using more established fossils, to make a convincing case for linking the Scottish and Welsh strata.[4]

This had been music to Murchison's ears. He was constantly on the lookout to expand 'Siluria', and for three successive summers during the early 1850s, while the row with Sedgwick rumbled from conference to conference, the two men had explored the hilly border country and confirmed to their satisfaction that the rocks were similar to those Murchison still regarded as his Lower Silurian strata in Wales. It had been a welcome distraction. The pair now planned to extend their collaboration to the northern Highlands, where the rocks were still poorly understood and therefore represented another promising opportunity to claim more territory.

The general consensus at the time (inasmuch as there was one) suggested that northern Scotland was made up of a series of bands of rock running roughly north–south. In the far west, passing up through the Western Isles, was a band of colourful crystalline rock known as Lewisian Red Gneiss: this was assumed to be ancient primary rock dating back to the creation of the Earth.[5] As it reached the coast of the mainland it gave way to a band of hard brown and red sedimentary sandstone known as Torridon, after a neighbouring sea loch. Some miles inland it was replaced by a third belt, a broad expanse of crystalline rock that stretched across the entire central region of the Highlands in a series of empty moors and bogs known as the 'flow country'. Finally, along the eastern coast of the Highlands there was another belt of red sandstone that dropped into the North Sea in a chain of spectacular red-brown cliffs.

The problem was how to fit these bands into the standard stratigraphic table. One view favoured by many (including by Murchison) was that the two areas of red sandstone were probably the same and, since the sandstone in the east contained similar fossils to those found in the Devonian Old Red Sandstone, the Torridon sandstone in the west was also assumed to be Devonian.[6] Inland, the broad expanse of crystalline rock – in reality a confused jumble of granite and metamorphosed quartzite, schists and gneiss – was thought to be older, though nobody was quite sure how old.[7] This meant that, in simplified terms, the geology of the Highlands resembled a sandwich: a central 'filling' of ancient rock sandwiched between two slices of younger Devonian Old Red Sandstone.

It was this central filling that had attracted Murchison. According to the stratigraphic table, if the bands of rock to the east and west were Devonian, then it was logical to assume the older rocks sitting below them would be Silurian, meaning that he might be able to claim not only the southern borders of Scotland for Siluria, but a huge swathe of the northern Highlands as well.

There were, however, several anomalies. On the northern coast of the central crystalline plug, near the remote village of Durness, an amateur geologist called Charles Peach had found some badly damaged fossils that appeared to be Devonian, which implied that at least some of these crystalline rocks might be as well. There were also several prominent sites

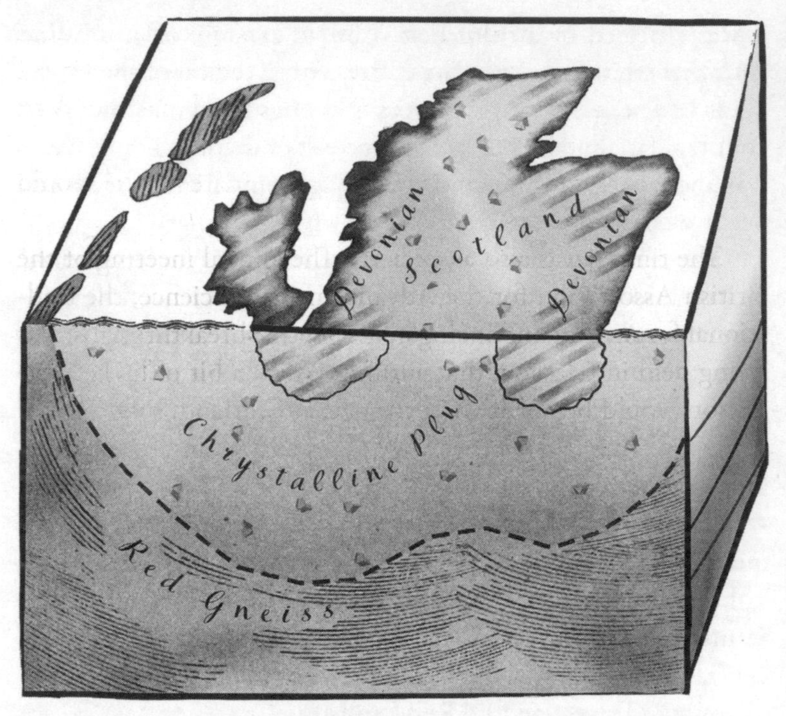

9.1 A cross-section through the rocks of northern Scotland, circa 1843, illustrating the 'sandwich' model. It shows two 'slices' of red sandstone running down the east and west coasts and a central 'filling' of crystalline rock.

across the region where these supposedly older crystalline rocks seemed to be sitting on top of the younger Old Red Sandstone – a stratigraphic impossibility unless the whole region had been turned upside down. Put together, these perplexing observations meant that the region's geology was generally regarded as a work-in-progress.

As a result, as Murchison and Nicol took the road west across the central Highlands, they probably had two slightly different goals in mind. Most immediately the region offered an opportunity to set aside the Welsh confusion and grab more territory for Siluria. But uncertainty about the exact identity of the strata suggested that the geology of Scotland cried out for a more careful and thoughtful investigation and, as a Scotsman, Murchison may have felt it peculiarly appropriate that he should be the man to do it.

The timing was also auspicious. The annual meeting of the British Association for the Advancement of Science, the traditional forum for announcing new scientific breakthroughs, was being held in Glasgow that autumn. With a bit of luck, Murchison would be able to visit northern Scotland, sort out the order and identity of the rocks and immediately present them for the approval of the scientific community.

The road west from Dingwall through the central Highlands climbs through a landscape of heather and peat moors, bereft of almost any sign of human occupation. Bog cotton and buttercups line the verges, small lochs fringed with bright-green grass and reeds reflect the colours of the sky.

Near the tiny settlement of Oykel Bridge the River Oykel breaks into a series of braids, and below the shallow brown water it is possible to see beds of crystalline rock that suggest the river is flowing over the central plug of schists and gneiss. Then gradually, like heads popping up on the horizon, the extraordinary red sandstone hills of the region, which is known as Assynt, appear on the western horizon, one after another: the bald dome of Canisp, the almost vertical twin peaks of Suilven and Cul Mor and, to the north, the treeless triple peaks of the Quinag massif – dozens of rusty-brown rocky peaks rising steeply out of the surrounding bogland.[8]

The track has passed from the central crystalline rocks to Torridon sandstone: the western arm of the Devonian sandwich.

It's a dramatic and unforgettable landscape. But to a geologist, what makes these mountains particularly remarkable is that on a clear day you can see how their red sandstone slopes are topped by pale-grey peaks of quartzite: one of those puzzling areas where apparently older crystalline rock, possibly dating back to the Silurian Period, appears to sit on top of younger Devonian sandstone. How could that be?

Murchison, however, showed little interest in this strange phenomenon. The two men pushed on with barely a pause, continuing north through a landscape of wide valleys and fast-flowing streams, past remote coastal settlements where it was reported that the inhabitants would come out to stare at the unusual sight of a passing carriage. Finally they reached the far northern coast, where a cluster of houses clung to the cliffs. This was Durness, where Peach had found the puzzling 'Devonian' fossils in what was widely suspected to be a band of limestone running through the Silurian rock. But once again Murchison showed little appetite for detailed fieldwork. The bright weather had vanished for several days; incessant rain and low clouds had obscured the surrounding hills. In the grey mist Murchison was feeling his age. 'All around gave note that my day was fast coming,' he wrote in his journal one evening, 'and that I had taken my farewell look at the Whiten and Far-out Heads' (the names of local headlands).[9]

The two men turned east, following a clifftop track along the far northern coast of Scotland, through the bleak moorland of the central schists: an area known as the Moine, after the Gaelic for 'bog' or 'moor'. Miles of dark waterlogged heather moors stretched in every direction. There was almost no sign of human life. '[My] heart was cheered by the sight of a road marked with cart-wheels, a thing which we had not seen for a long time,' wrote the diarist James Boswell, passing through similar country. 'It gave us a pleasure similar to that which a traveller feels, when, whilst wandering on what he fears is a desert island, he perceives the print of human feet.'[10] Small settlements of derelict cottages added to the desolation – bleak reminders of the recent Clearances.

Finally, amid the roar of swollen streams running off the moors, Murchison and Nicol reached the eastern coast of Scotland, where brown sandstone cliffs dropped steeply into the North Sea. They had arrived in the eastern arm of the Devonian 'sandwich', where the fossil fish Holoptychius had been found in the cliffs: the only rocks in the Highlands that could be definitively correlated with the wider stratigraphic table.

Back in Glasgow at the British Association conference, there was the almost obligatory stand-off with Sedgwick, who had recently published a paper on early British fossils[11] in which he had criticised Murchison with 'a vehemence of language very characteristic but regretted by all who admired and respected both disputants.'[12] As a result, delegates were well primed for trouble. 'It was impossible to conceal the fact,' wrote Murchison's biographer later, 'that the difference remained no longer a mere scientific one, but had descended, as such differences almost inevitably do at last, into personal estrangement.' At one stage the animosity turned to farce when, taking off his greatcoat to speak, Sedgwick noticed smiles among the audience and joked, 'Oh, I'm not going to fight him!' – at which point the audience broke into uneasy applause.[13] Later Murchison reported on his recent trip, confirming his belief in the 'sandwich' model of the Highland's geology, but making no serious attempt to address the anomalies that had challenged this interpretation.

It had been a hurried and superficial trip that achieved little, except to distract attention from the dispute with Sedgwick, and Murchison wisely held back from making too many claims for Siluria in the Highlands. 'The results of this tour were not very convincing,' his biographer Archibald Geikie would later write. 'The rocks of the north-west of Scotland presented a very curious and interesting problem which could not be solved without more and better fossils, and further extended examination of the ground.'[14]

That winter, ruminating on the problem in his university rooms in Aberdeen, James Nicol resolved to make a return visit: the journey with Murchison had left too many unanswered questions.[15] One series of rocks in particular bothered him. For years, geologists had noticed a prominent line of inland cliffs and rocky outcrops running up through north-west Scotland, from the small fishing port of Ullapool on the western coast to the north coast at Loch Durness, roughly following the western edge of the central crystalline plug that Murchison was hoping to claim for Siluria. Here sections of rock had been heaved up, leaving bare cliffs where a cross-section of the underlying strata was clearly visible. This seemed a good place to conduct a more detailed study of the rocks that lay immediately below and around the crystalline core.

During the summer of 1856 Nicol travelled west from Aberdeen to Ullapool. On the outskirts of the port, where herring trawlers clustered along the waterfront and nets were hung to dry, he picked up the southern end of the cliffs and followed them north to Durness. At various points along the way the cliffs stood proud against the surrounding countryside and the lines of strata were prominently displayed.[16] Time after time they revealed a similar pattern. Sitting at the bottom was a basement of Red Gneiss. Immediately above it was a prominent break or unconformity in the strata, where the rock record had vanished. Then came red Torridon sandstone characteristic of the Assynt, followed by a series of narrow interwoven bands of limestone and quartz. And then, closely related to them but sitting on the top, a cap of the so-called Moine schists and gneiss. This proved conclusively that the central plug of crystalline rock sat on top of the Torridon sandstone and should, logically, be younger.

It was immediately obvious to Nicol that Murchison's ordering of the rocks the previous summer had been wrong. Instead of two bands of Devonian sandstone bookending a central plug of older Silurian rock, the new ordering suggested that, if the red Torridon sandstone was Devonian, then the distinctive limestone-quartz rock sequence sitting on top of it should be Carboniferous. This in turn implied that the closely associated central plug of crystalline rocks, the Moine schists and gneiss, were also Carboniferous. At the very least the limestone-quartz sequence occupied some kind of intermediate position between the Devonian and Carboniferous, which made it entirely possible that the fossils Charles Peach had found in the limestone cliffs at Durness were indeed Devonian.

This gave Nicol a completely new vision of the structure of the region's rocks. Where Murchison had seen the central Highlands as a Silurian anticline, Nicol now envisioned them as a Carboniferous trough – a direct challenge not only to Murchison's reading of the rocks the year before, but also to his hopes of spreading Siluria to northern Scotland. This must have caused Nicol, a man regarded by friends as shy and even rather diffident, an uneasy moment.[17] He owed much to Murchison, including his position at the University of Aberdeen. Nicol would have had little desire to oppose somebody he regarded as a friend, as well as one of the most influential and internationally recognised authorities on the Earth's early rocks. Perhaps he hoped Murchison would receive his new speculation in the spirit of open enquiry.

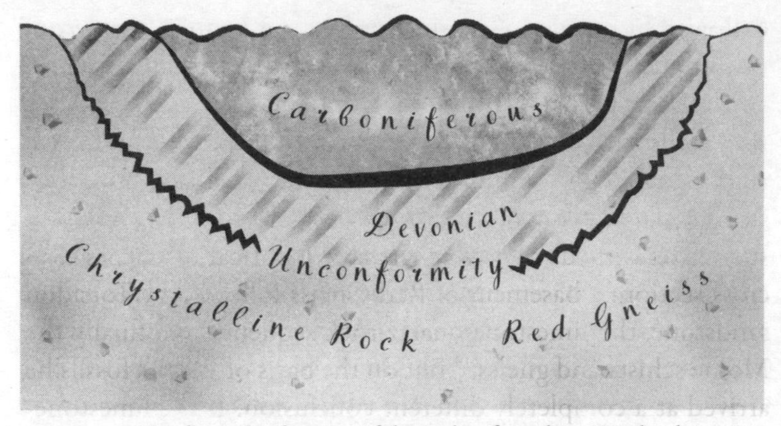

9.2. Nicol's 1856 redrawing of the rocks of northern Scotland, compared with Murchison's 'sandwich' model.

Unfortunately for Nicol, there had in the meantime been developments in London. John Salter, now the Geological Survey's official palaeontologist, had finally examined Peach's Durness fossils in detail and decided they probably weren't Devonian after all. Neither were they Carboniferous. Salter had compared them with fossils from elsewhere in Britain and abroad, and concluded that at least five of them could well be Silurian, and in particular from Murchison's Lower Silurian Llandeilo or Caradoc rocks.[18] Murchison, still basking in his success in Russia and not a man to take lightly challenges to his authority – even those made in the spirit of open enquiry – seized on the new evidence. It undermined Nicol's findings and suggested that if the Durness Limestone was Lower Silurian, then there was no reason why the closely associated Moine schists should not also date from that period. To Murchison, that settled the matter.

Others were not so sure.

*

Over the next two years Murchison and Nicol made a number of rival trips to northern Scotland. Murchison revisited the northern coast and a large sea loch called Eriboll, where bare hills rise steeply from the shore and on a bright day bands of quartz running through the slopes sparkle in the sunlight. At several sites he drew illustrations of the rock strata, which showed a succession of beds that were virtually identical to Nicol's 1856 cross-section: a basement of Red Gneiss followed by Torridon sandstone, the limestone-quartz rock sequence and finally the Moine schists and gneiss.[19] But on the basis of Peach's fossils he arrived at a completely different conclusion. If the limestone-quartz series was Lower Silurian, then according to the standard stratigraphic table the Torridon sandstones below it – until now regarded as Devonian – might be Cambrian. This would suggest that the Moine schists and gneiss above might be highly metamorphosed Upper Silurian limestones heated and transformed by local volcanic activity.

To Murchison, this suggested yet another way to understand the rocks of northern Scotland. Out went the old idea of a Devonian sandwich with a filling of older crystalline rocks, and out went Nicol's model of a Red Gneiss-Devonian–Carboniferous trough. In their place Murchison proposed a smooth succession of strata running across northern Scotland, starting with the Primary rocks of the Red Gneiss on the west coast, continuing with a smooth and unbroken sequence of Cambrian and mainly Silurian rocks running up the west coast and across the central region, and ending with a band of Devonian sandstone, the only rocks with an absolutely clear fossil identity, on the east coast.

Unlike Nicol's model, which implied large disturbances in the deposition of the rocks and a massive gap in the rock record between the Primary Red Gneiss and the Devonian sandstone where the Cambrian and Silurian periods had

9.3 Murchison's 'grand revolution' in the understanding of the geology of the Highlands was markedly different from Nicol's 1856 section.

completely vanished, Murchison envisaged a history of smooth and unbroken deposition over a period of hundreds of millions of years. Nicol saw Scotland as a region of highly disturbed geology. Murchison saw it as the opposite: a place where the complete early history of the Earth was laid out and clearly displayed. This brought the geology of northern Scotland neatly into line with that of the rest of Britain and, in passing, gave much of the region to Siluria.

Murchison called it his 'grand revolution' in the understanding of the geology of the Highlands.[20] As Archibald Geikie would later describe it in more flowery prose, Murchison – 'map in hand, and tracing out in detail the boundaries of the rocks' – had proved 'the great mass of the crystalline rocks of the Highlands as altered strata of Lower Silurian age'.

Nicol, meanwhile, also retraced his tracks. One feature of the region's stratigraphy continued to trouble him: at a number of sites along the line of outcrops and cliffs he had noticed that the highly metamorphosed Moine schists and gneiss were sitting more or less directly on top of beds of limestone, quartz and Old Red Sandstone. And these seemed completely unaffected by the heat and pressure of metamorphism.[21] Nicol simply couldn't imagine a scenario where this might happen: he knew that violent energies of metamorphism normally left a clearly visible change in the colour and chemical composition of adjoining rocks: the process is known as 'contact metamorphism', and is frequently indicated by dark spots. But Nicol could find absolutely no evidence of this spotting.

There was, however, evidence of a different kind. At various locations he found clusters of reddish crystals of minerals such as feldspar and serpentine. These are common at sites where molten magma has forced its way up through the Earth's crust to the surface, cooling rapidly and forming clusters of small mineral polyhedrons. But these were much larger crystals – a characteristic of minerals that have cooled more slowly over many millions of years. To Nicol, the absence of contact metamorphism, combined with the presence of clusters of large crystals, suggested a different story again.

Suppose the central plug of schists and gneiss was neither Carboniferous nor Silurian, but dated from a much earlier period when these rocks had been heated and metamorphosed deep in the Earth's crust and left to cool slowly? Millions of years later, movements in the Earth could have seen a huge block of this metamorphic rock pushed up and over younger, unmetamorphosed strata until it reached its current position. This would mean the limestone-quartz rock sequence represented the jumble and chaos of an enormous fault line where ancient metamorphic rocks – possibly Primary rock of the same age as the Western Red Gneiss – had been pushed up and eastwards in a giant sliding motion, and ended up sitting on top of younger sedimentary strata of the Torridon sandstone.[22] Instead

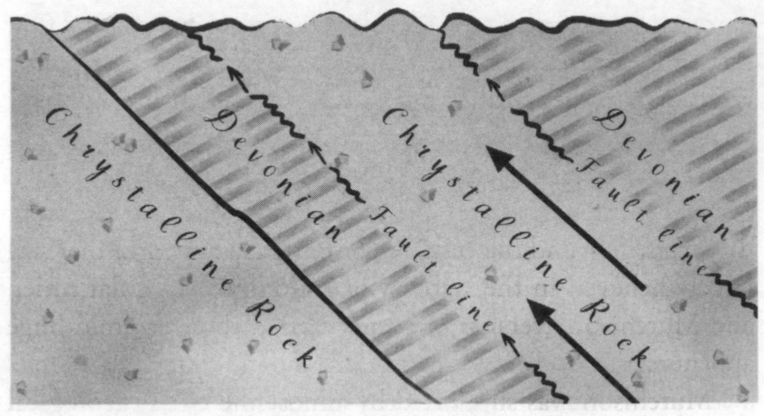

9.4 Nicol's modified version of the rocks of northern Scotland, suggesting a huge fault line running through the country where rocks moving east to west had been pushed up and over younger strata.

of Murchison's unbroken succession of strata running from west to east across the country, Nicol saw a region of upheaval with a broad fault line running up through northern Scotland from Ullapool to Durness.

A showdown was inevitable. It came in the autumn of 1859, at a poignant meeting of the British Association in Nicol's home town of Aberdeen. It was said to have been attended by around 2,000 people, the biggest gathering of scientists ever held in northern Scotland. Murchison, recently appointed director of the Geological Survey in the wake of De la Beche's death, was the celebrity speaker.[23] During a packed evening session in the University's New Hall, he reiterated his essentially 'simple' structure of the Highlands. His presentation was said to have been an uncharacteristically faltering and hesitant one, which failed to address any of the complexities of the region. Instead Murchison fell back on the widely held, almost unconscious assumption that broadly horizontal beds of rock, in a regular chronological pattern, were the 'natural order' of the Earth and that major disturbances, such as might occur during a volcanic eruption, were never more than localised phenomena.

Nicol, the leading geologist in the city and heavily involved in organising the event, outlined his alternative vision. But it posited such a huge disturbance in the Earth's surface that it was simply beyond the comprehension of an audience that still largely believed in the stability of a God-given regular order, and Murchison's version of events carried the day, amid huge applause.[24]

Murchison was supported by almost the entire geological establishment, including, as it happened, Sedgwick. The enormously influential populariser of geology, Charles Lyell, proposed the vote of thanks, and John Phillips, now promoted to professor of geology at Oxford University, talked of the 'high esteem' in which Sir Roderick Murchison was held 'over half the globe', while the Royal Society of Edinburgh presented him with its Brisbane Medal for scientific research. The *Aberdeen Journal* spoke for many when it reported: 'so plain and irresistible were the evidence presented in a number of transverse sections' that the only wonder was 'that any scepticism should have prevailed as to the order of succession'. Nicol's lecture, by contrast, received hardly a mention, with the *Aberdeen Journal* briefly reporting, with damning condescension, that he had given 'an able and interesting notice'.[25] Murchison returned to London bathed in glory, stopping off to pay a visit to Queen Victoria and the royal household at Balmoral, where he was treated to several hours of Highland Games.

In some respects the 'Highland Controversy', as it became known, looked like a rerun of the Welsh feud between Murchison and Sedgwick. There was one important difference, however. Murchison had none of the respect for Nicol that he had held for Sedgwick, and no compunction about insulting his Scottish compatriot. When Nicol brought a new paper to the December 1860 meeting of the Geological Society in London, we are told that Murchison couldn't bring himself to read an advance copy. And on another occasion he said to a colleague:

> will you condescend to give him a good back-hander or two ... he is a bitter biter – and I rue the day that I had him with me & pushed [?] him etc. [?] through the Highlands. He has not only what the French call 'Mauvais fois' by the way; but he slyly went by himself the next year & then tried to turn me over & get all the credit himself. I wish you would hammer him.[26]

These were words Murchison would never have used against Sedgwick, and they carried considerable weight. In his new capacity as director of the Geological Survey, Murchison wielded enormous power and influence. In the fierce competition for the few paid posts in British geology, a significant number of young geologists owed their position and prospects to his patronage, and Murchison had no hesitation in taking advantage of it. Over the following months he worked hard at undermining the 'stubborn Aberdonian' and persuading the young scientists of the Geological Survey that his only supporters were 'some old mineralogical fogies at the Scottish school'.[27]

Nicol wilted under the onslaught. There is no record of his reaction, and he continued to spend his summers quietly charting the rocks of Scotland, occasionally 'accompanied by an old geological friend to whom he could communicate' his thoughts.[28] But the Highland Controversy marked the beginning of his isolation from the leading circles of British geology.

9.5 James Nicol in later life.

His career stalled and his only noteworthy publication in later life was a short update to his small book on *The Geology and Scenery of the North of Scotland*.[29] Gradually he fell 'behind the crowd of younger aspirants to scientific reputation', in the words of one observer, and became 'less widely known than his ... qualities deserved'.[30] A picture of Nicol in later life shows a heavily bearded, severe and unostentatious man gazing out at the world with a shy, slightly guarded look.

Yet Murchison's ruthless determination to push through his extended Silurian System, despite the uncertain and confusing picture in both Wales and Scotland, was not universally welcomed. There was a sense among younger colleagues that he was growing tired and slapdash in his fieldwork: more interested in evidence that fitted his picture of the world than in any serious attempt to discover the truth. They worried that, in the rush for answers, important clues were being missed or overlooked, and that the story of the rocks might be more subtle and complex than anybody suspected.

'What an extraordinary man Sir R is', wrote one member of the British Geological Survey to a colleague. 'I think he would sacrifice truth and everything else [rather] than his first ideas of the Silurian System should be in the least controverted ... It is very poor for him to make to say that all his original ideas were quite correct in every subdivision in spite of our labours showing the contrary.'[31] Another wrote of Murchison being 'so full of self-love or rather self-esteem and love of glory that he would like to play the part of Creator'.[32]

What was more, many younger geologists had trained under Sedgwick in Cambridge and were unhappy to see Murchison disparage his findings and achievements. The Geological Survey's palaeontologist John Salter, who had identified the Durness fossils as probably Silurian, wrote to Sedgwick at around this time:

> There are some of us owe a great deal to you, and never think of you without true regard. Here am I running along the South Welsh boundary – entrusted with the whole fossil work of three kingdoms! (for at least a while.) And who took me into the field, and taught me to know ... [a bee ?] from a bull's foot, but your-self.[33]

Joseph Beete Jukes, another member of the Geological Survey who had trained under Sedgwick in Cambridge, wrote to him in a similar vein, addressing him as 'My dear father' and concluding, 'Upon the course of my life your influence has been as great as beneficial and while it lasts you will always be most gratefully remembered.'[34] (He would later write, a little unfairly, of Murchison that 'all he has done has been by the strength of his purse, buying the brains of other men'.[35])

*

Murchison, meanwhile, was now in his late sixties and increasingly preoccupied with old age. While he continued to fight for Siluria, he was also growing anxious for a reconciliation with a man he still regarded, in the quiet moments of the night, as a friend and mentor. When the second edition of his new 'big book', *Siluria*, had been published in late 1859 he had sent Sedgwick a copy and been pained to get no acknowledgement. A week later he wrote to Sedgwick again: 'Time rolls on, and as we passed many a happy day together, I trust that you will have some gratification in turning to these pages, particularly those relating to the Highlands of Scotland.' But it was his signing off that conveyed his real feelings: 'Clinging to the hope that the only bitter sorrow I have experienced in my scientific life may pass away, and that your old friendly feelings towards me may return.'[36]

They never did. Sedgwick remained implacable to the end. At the 1861 annual meeting of the British Association in Manchester, Murchison announced that the Geological Survey had adopted his 'extended Silurian' version of the Welsh rocks. Of course they have, Sedgwick is said to have snapped, 'they have been badly instructed and badly led, and it is no wonder that they have adopted your erroneous views.'[37] The following year, at the 1862 annual conference held in Cambridge, Sedgwick accused Murchison of being 'false in word and deed' and of committing 'one of the greatest and most mischievous blunders in the history of English geology, a direct and deliberate untruth.'[38] It was from fear of just such an outburst that Murchison had declined to attend. 'I believe you have acted contemptuously, unjustly and falsely,' Sedgwick later wrote to him, accusing him of being afflicted with a 'self-love' that made him 'quite oblivious of those important duties which belong to the courtesies of life'.[39] He felt very unfairly treated and continued to call for an apology and retraction of Murchison's claims to an extended Siluria in Wales.

Murchison confided to his journal afterwards, 'I am much out of health, and certainly all the worse for the coolness and chill with which my former friend Sedgwick received every effort I made to be on our old terms with him. I could not have believed that a man whom I had always looked on as the most hearty and generous of beings should have turned upon me with unforgiving tenacity.'[40] He later wrote to a colleague, 'if I did not feel that he had irrevocably made up his mind to be alienated from me I would still make every effort in my power to win back his friendship ... I am at a loss to know why in the last year he has become so morose and unforgiving.'[41]

Sedgwick was now in his seventies. He was frequently ill and had little appetite for sustained work. He still gave lectures, but his style had become frustratingly, if amusingly, discursive. 'The sight of some fossil on the table, or some diagram on the wall would recall the occasion on which he found the one or noted down the memoranda for the other,' wrote a former student:

> and off he would go, recounting the 'glorious time we had together' with an exuberance of cheery reminiscence, a vivid picturesqueness of description and a homely raciness of language, the impression of which remains with me still. Then, after an apologetic 'But this is all by the way gentlemen, this is all by the way,' he would abruptly fall on his assistant with, 'where is that fossil? Eh? what did you say? Oh! Ah! Yes!' And then launch out into a description of the animal of whose body the fossil in question had once formed.[42]

Eight years later, in 1869, Murchison's wife, Charlotte, died. Sedgwick finally wrote to him, but it was a stiff letter. Gone was the friendly and informal 'Dear M' of the past. 'Dear Sir Roderick Murchison,' the letter began formally:

> I did not wish to intrude myself on your sorrows too soon. Indeed, such has been my life of solitude for the last two months, that incidents of the greatest interest to my heart have more than once passed away for a full week or ten days before they reached me. You will, I know, believe me when I say that the first news of your beloved wife's death filled me with very deep sorrow.[43]

Murchison replied, reiterating his desire for a reconciliation. 'It would give me the sincerest satisfaction, if any explanation which I can possibly give would be acceptable to you, and put an end to an estrangement which for more than twenty years I have never ceased to lament.'

Sedgwick almost certainly did not reply. Time, anyway, was running out. In October 1871 Murchison himself died. He was buried in London's Brompton Cemetery with the sort of pomp and ceremony he had enjoyed and cherished all his life. 'The Queen and the Prince of Wales testified their respect by sending their carriages to join the funeral procession. Among those who walked bareheaded behind the bier the most conspicuous form was that of the prime minister, Mr Gladstone,' recorded the faithful Archibald Geikie, who was almost certainly there himself.[44]

Sedgwick didn't live much longer. Fifteen months later, in January 1873, he died in his rooms in Trinity College, Cambridge. His memorial stone lies in the floor of the antechapel at Trinity College next to his lifelong friend and colleague William

Whewell, who had worked hard to damp down the fire of the conflict with Murchison.

Many people saw this as the end of an era in British geology. Both Henry De la Beche and George Greenough had died in 1855. John Phillips, who had advised both Sedgwick and Murchison, would pass on a year after the former. Charles Lyell, the influential theoretician and populariser of geology, died two years later. Despite the enormous achievements of this pioneering generation of Victorian geologists, the identification and stratigraphic order of the world's older rocks had reached an impasse, frozen in confusion and uncertainty.

In Wales the boundary between the Cambrian and Silurian periods was still in dispute. In Scotland Murchison's attempts to grab most of the Highlands for Siluria had left niggling doubts. It was time for a new generation of geologists, 'a younger race, sometimes of great vigour',[45] to look with fresh eyes on the problem. In the forefront was a thirty-year-old provincial school teacher and amateur geologist who had just moved to the gently rolling hills of the Scottish borders. His name was Charles Lapworth.

Chapter Ten

THE MISSING KEY: DISCOVERY OF THE ORDOVICIAN, 1864–84

The small town of Galashiels sits in the steep-sided Gala Valley, amid the rolling hills of the Scottish borders. It was once a hard-working factory town: more than twenty mills used the power of the River Gala's soft water to wash the wool and power the looms of a booming textile industry. In the 1980s, however, one factory after another fell victim to cheap imports and the Victorian mills were torn down, ripping the heart out of the old high street. Young people moved away; pound shops moved in. Today only the occasional store still sells tweed and kilts to passing tourists. This chronicle of success and failure is recounted in the local museum and on websites devoted to the area's past.

But one important story is missing: there is nothing about the life and times of one of the town's most influential residents, a man who lived and worked in Galashiels for eleven formative years in the 1860s and 1870s. And yet, without him, the puzzle of the Greywacke might never have been solved.

Charles Lapworth, known to all as Chas, was a recently qualified teacher in his early twenties when he moved from his native Oxfordshire to the Scottish borders in 1864. He had come to teach English and art at one of the town's church schools, though it is said he was attracted to the region's wild open hills by the novels of Walter Scott, who had recently built himself a baronial manor-house nearby.[1]

Lapworth was bright, and curious about the natural world, and as he explored the countryside around the town, perhaps tracing places and events mentioned in Scott's novels, he became interested in the region's geology. He struck up friendships with local geologists from Edinburgh, and possibly ran into a team of surveyors from the Geological Survey who were mapping the surrounding hills.[2]

Ten years earlier, in the first flush of their cooperation, Murchison and Nicol had explored the region and concluded that the rocks were composed of a succession of gently sloping beds of dark shales – a highly compressed form of mud – and gritty sandstones. There were known to be faults and folds that complicated the picture, but generally the rocks were regarded as relatively undisturbed. A frustratingly sparse fossil record made identifying their age hard, but their gritty grey composition and appearance suggested they were probably similar to the Caradoc Sandstones and Upper Llandeilo shales of the Welsh Marches, so Murchison and Nicol had concluded that they were Lower Silurian. The newly arrived Geological Survey, now under Murchison's direction, reached a similar conclusion. Yet, as so often, puzzles remained.

If you added together the width of the dozens of bands of shale and sandstone, their total depth came to some 26,000 feet, which made the Lower Silurian rocks of the Southern Uplands about the same thickness as the entire Silurian Period elsewhere in the world. This seemed odd. To add to the uncertainty, such fossils as did exist consisted primarily of the tiny, little-understood graptolites. In short, the identification with Murchison's Lower Silurian Caradoc and Llandeilo rocks was 'confessedly vague and generalised', in the words of one prominent Scottish surveyor, and rested 'only on a combination of probabilities'.[3] As he wandered the hills, Lapworth may have wondered whether the bitter controversy over the boundary between the Cambrian and Silurian periods was playing out beneath his feet.

The hills around Galashiels are round and smooth, worn down by erosion, and there are few natural outcrops of bare rock. But in the late nineteenth century the booming industrial towns of the region needed building stone and transport routes and the hills had been sliced open by quarries and railway lines, exposing the underlying strata. (Such was the allure of the new railway cuttings for geologists that it had led to at least one death, when an unwary explorer was 'cut in two pieces at the mouth of a tunnel'.[4])

One of the more prominent sites was a long, deep cutting near the tiny hamlet of Thornylee, where the newly constructed railway to Edinburgh carved through the hills of the Tweed Valley. Chas Lapworth and his friend James Wilson, a local journalist and later editor of *The Border Advertiser*, became frequent visitors to this wall of freshly exposed rock. They soon found that it didn't corroborate Murchison and Nicol's simple model of the region's geology at all. At dozens of points along the cutting there were signs of sharp, abrupt folds and major faults in the strata that belied the idea of gently sloping beds of undisturbed rock. The strata, often paper-thin, ran at crazy angles, dipping, diving and sometimes completely inverted, making it virtually impossible to distinguish one bed from another or work out the relationship between them. Furthermore the fossil record, as Nicol had discovered, was extremely thin. Any attempt 'to ascertain their interrelationship by lithological and stratigraphical evidence has soon to be abandoned as hopeless,' wrote Lapworth. 'Their separation by zoological characters appears to be quite as desperate; for in many localities every trace of their former prolific fauna has been obliterated; in others only one or two fragmentary forms are obtainable, and these are limited to a few inches of thickness of the less altered zones.'[5] It was the first indication that Murchison and Nicol had significantly underestimated the complexity of the region.

Lapworth, still in his early twenties, must have wondered how to proceed. Probably more in hope than conviction, he turned to the only plentiful fossils he could see: the graptolites. At first sight they looked more like random scribbles scratched in the rock than the remains of living creatures; indeed, this likeness to scribbles had given rise to their name: from *graptos*, the Greek for 'written', and *lithos*, meaning 'rock'. Sometimes they resembled tiny spirals, at other times Y-shaped branching stems, saw-blades or serrated fish hooks – peculiar markings that nobody had ever made much sense of. But in the puzzling rocks of the Scottish borders Lapworth could find little else to work with, and so in the mid-1860s he began systematically to collect and catalogue the graptolites, remarking that 'these peculiar creatures have never yet received the attention they deserve'.[6]

It was an extraordinarily testing and painstaking process. Most graptolites are small: hard to find and easy to miss. To help him collect them Lapworth is said to have devised, with the help of his wife Janet, what he called his 'geological waist coat', a jerkin with a series of pockets down each of the front panels into which he would carefully place his often minute and delicate specimens.[7] Gradually Lapworth and Wilson amassed a collection of at least forty distinct species. Many were almost indistinguishable from each other and had never previously been identified.

But the fossils alone were only part of the story. To use them

10.1 Strange and ambiguous markings left by graptolites.

as markers of geological periods it was crucial to know exactly which species came from which bands of rock. As Lapworth and Wilson explored the quarries, cuttings and stream beds of the hills around Galashiels, this became a major issue. The more they looked, the more complex and disturbed the strata appeared: infuriatingly fine layers of rock, twisting, plunging, merging and sometimes completely vanishing. It forced them to literally crawl, almost inch by inch, over every section of exposed rock, painstakingly tracing out the strata as best they could and logging the exact location of every fossil find. At times they must have wondered whether they had set themselves an impossible task. Yet slowly they uncovered a remarkable story.

One of their most fruitful sites was a small gorge just off the Galashiels-to-Moffat road. On a summer morning when I drove to it there were splashes of purple heather and small stands of mountain ash heavy with red berries amid the green vastness of the hills. Slowly the road winds out of the Tweed Valley and descends into the valley of the Moffat Water, past a remote whitewashed shepherd's cottage at a place called Birkhill. Just to the north, across a grassy slope, lies a gash in the hillside where the upper tributaries of the Moffat have cut into the rock, exposing band after band of black and grey shale and sandstone. Small streams and waterfalls rush down the steep sides. It is known as Dobb's Linn,[8] and is one of the few extensive sections of naturally exposed rock in southern Scotland. Lapworth must have fallen on it like a dying man stumbling on an oasis in the desert.

Here, over a period of five to six summers during the 1870s, Lapworth, probably supported by his wife and certainly by Wilson, spent entire months camped out in the shepherd's cottage, mapping in minute detail the thin, elusive bands of mudstone strata, breaking off shards of razor-sharp rock. Many bore the faint imprint of graptolites, and Lapworth would carefully place them in the appropriate pocket of his waistcoat. He was patient, careful and methodical and is said to have possessed a 'sharp eye for lithological variation and a vivid memory for lithological type'.[9]

Gradually, over the painstaking months, a pattern emerged. Each band of shale, sometimes only inches thick, appeared to be composed of distinct layers or zones. To all but the most discerning eye they were virtually indistinguishable, but over several years Lapworth found he could identify five different types in each band of shale: a top zone of dark-grey muddy siltstone and black shales; an almost imperceptibly different zone of grey mudstone; a zone of dark-grey and black shales with cherty (that is, flinty) ribs, visibly very different from the first two zones; a second zone of black cherty shales; and, finally, a third zone of black cherty shales separated from the zone above by a thin cap of mud and siltstone.

This was impressively meticulous work, but it was his next step that represented the real breakthrough. At the shepherd's cottage at Birkhill, Lapworth installed a five-drawer cabinet, each drawer corresponding to one of the zones in the shale. Every evening, after a day collecting fossils, he would systematically transfer them from his waistcoat to the appropriate drawer in the chest. Over time he found he could correlate, with great accuracy, each graptolite with its zone in the shale. He described the procedure: 'We place all the graptolites we procure from the highest grey shales in the first set [of drawers], those from the second set of beds in the second set of drawers, and so on. In our spare time of an evening we name and identify these graptolites but we are always careful to replace them in the proper drawer.'[10]

It was in the process of doing this that Lapworth made a remarkable discovery:

> we find that our drawers contain three distinct sets of graptolites. The two upper sets of rocks contain one set of graptolites. Those answering to the two middle sets of shales contain a second set and those answering to the flinty beds at the bottom of the section contain a third set. None of these species of graptolites found in one set passes into another.

In other words, each band of shale was composed of three distinct graptolite zones, as follows:

1. The top two bands of mudstone, siltstone and black shale. He called these the Birkhill Shales, after the Birkhill shepherd's cottage.

2 The middle two bands of grey and black cherty shales. He called this the Hartfell Shale, after a hill that rises to the east of Dobb's Linn.
3 The bottom band of black cherty shale. He called this the Glenkiln Shale, after a nearby river valley.

It was the first time anybody had used graptolites as fossil markers, and it meant that large areas of Greywacke rock that lacked a more obvious fossil fingerprint could finally be pinned down. Nobody had ever explored rocks in such minute detail before. Walking into Dobb's Linn today, following a faint footpath beaten down by the feet of visiting geologists, I have a sense of walking on hallowed ground: *this* was where Lapworth and Wilson revolutionised our ability to read the ancient rocks.

Over the next couple of years Lapworth methodically explored the surrounding hills, where fast-flowing streams had scooped out banks and small ravines, putting his findings to the test. He travelled east to the North Sea coast of Berwickshire, and west to the Ayrshire shores of the Irish Sea. At each location he found exposed beds of rock composed of similar bands of shale that contained, in greater or lesser detail, the same three graptolite zones he had identified at Dobb's Linn.

Finally, in 1878, ten years after they had begun their research, Lapworth and Wilson pulled their findings together in a seminal article they called the 'Moffat Series'.[11] It drew a number of momentous conclusions. If each shale band contained exactly the same graptolite fauna, in exactly the same pattern, then it almost certainly followed that far from there being a series of different bands of shale, the rocks were composed of the very same band, faulted and doubled back time and again, to create the illusion of numerous bands. Indeed, Lapworth and Wilson's minute examination of the zones within each band of shale showed all too clearly that many had been inverted and were now lying upside down.

Here was confirmation, if any was needed, of the heavily disturbed nature of the rocks of the region. Equally importantly, it meant that the apparently enormous and puzzling depth of what Lapworth still regarded as Lower Silurian rocks was an illusion. Far from being some 26,000 feet deep, as Murchison and Nicol (and the Geological Survey) had maintained, Lapworth and Wilson had shown that in reality it was the same band of Lower Silurian rock doubled over, time and again, to give the illusion of depth. The actual depth was closer to 300 feet: a measurement that brought them far more in line with Lower Silurian rocks elsewhere in the world.

The Moffat Paper established the unique value of graptolites as fossil markers. Lapworth showed that these curious marine organisms had evolved rapidly over relatively short periods of time, making them extremely sensitive and accurate markers of the development of the Earth's rocks. Moreover, their habit of drifting on ocean currents far and wide meant they could be used as markers across huge areas of the globe. It turned them from an obscure family of puzzling and largely ignored fossils into a vital source of data for mapping the Earth's early rocks. 'No other group of creatures, except ammonites, is a better guide to the correlation of strata in different regions far removed from one another,' wrote one later commentator.[12] A new form of fieldwork, 'micro' fieldwork, or 'graptolite zonation' as Lapworth preferred to call it, was born. It would, Lapworth predicted, 'effect in the history of geological research a revolution as great and an advance as rapid as those brought about by the use of the microscope in the history of biology'.[13]

All that remained was for Lapworth to correlate his three zones with the wider stratigraphic table. To do this, he chose

Murchison's widely accepted 'extended Silurian System' and, on the basis of the lithology and fossils of his zones, suggested they represented the Lower Silurian Caradoc/Bala and Llandeilo beds.[14]

It was an extraordinary set of findings, all the more remarkable for being conducted among some of the most complex and disturbed geology in Britain. 'There was not a single straightforward section, among traps and pitfalls innumerable, and where the rocks were as often on their heads as on their heels,' Lapworth's friend and fellow geologist William Watts would later write.[15]

There is no record of whether any word of these extraordinary developments ever reached the ears of Sedgwick or Murchison, but by the time the Moffat Paper was published, both men had been dead for several years. Even if they had read it, it is doubtful whether they would have been able to climb out of their respective redoubts and make anything of it.

Certainly much of the geological establishment was profoundly reluctant to revisit the controversy. The Moffat Paper went down badly with the Geological Survey, which had yet again failed to grasp the region's geological complexity and now faced the prospect of having to rethink its entire understanding of Scotland's rocks. But the wider geological community was impressed, and the following year Lapworth was awarded funds from the Murchison Geological Fund to complete his survey of southern Scotland. He must have enjoyed the irony; and, some twelve months later, in 1879, he took the next logical step.

In the words of his friend William Watts, he 'tactfully and modestly' weighed into the dispute between Sedgwick and Murchison, lining up his three graptolite zones with the disputed boundary between the Cambrian and Silurian Systems. The outcome was explosive.

At the top of his zonal table he announced there was a complete break in the fossil record between the Birkhill and Hardfell Shales. Between them, Lapworth wrote:

> a sudden and most extraordinary change takes place in the genera and species of the fossils of the Moffat Series. Scarcely a single form of those collected … in the [Upper Birkhill] strata already described is met within any of the beds below this line … Here, therefore, is the chief palaeontological break in the succession, and here, consequently, are we compelled to draw our main line of demarcation in any natural classification of the Moffat Series.[16]

Indeed, the break in the fauna was so significant that Lapworth concluded the Birkhill strata must belong to an entirely different geological period from the Hartfell and Glenkiln Shales, not just a different subdivision of the Silurian. In short, here again was the break or unconformity at the heart of Murchison's Silurian System that Sedgwick had already noted.

MURCHISON'S STRATIGRAPHY		LAPWORTH'S ZONES
UPPER SILURIAN	Ludlow	
	Wenlock	
LOWER SILURIAN	Caradoc/Bala	Birkhill Shale
		Hartfell Shale
	Llandeilo	Glenkiln Shale
CAMBRIAN		

10.2 Fitting Lapworth's graptolite zones into a simplified version of Murchison's extended Silurian System, circa 1878.

But this was more than just a rerun of the old argument about where to draw the line between the Cambrian and Silurian Systems. Lapworth argued that the Hartfell and Glenkiln Shales shared a sufficient number of fossil species to be bracketed together, but that below them there was another important change in the fossil record, where an entirely different graptolite fauna appeared. This meant the Hartfell and Glenkiln Shales shared a unique fossil fauna that was entirely different from the rocks above and below them. Here at last was why Sedgwick and Murchison had been unable to resolve the boundary between the Cambrian and Silurian: they had missed the fact that there was an entirely different, third geological period separating them.

It was nothing less than a revolution. Lapworth called it the 'Ordovician', after the last Romano-British tribe to inhabit north-west Wales. 'Every geologist will at last be driven to the same conclusion,' he wrote, 'that Nature has distributed our lower Palaeozoic rocks in three sub-equal systems, and that history, circumstances, and geological convenience have so arranged matters that the title here proposed for the central system is the only one possible.'[17]

Today we know that the Ordovician represents a period when marine animals rapidly increased and diversified and that, at the end of it, 443 million years ago, a major glaciation event killed some 85 per cent of all life, the third-biggest extinction in the history of the Earth, before the Silurian period kicked in.[18] The Ordovician took a large bite out of Murchison's extended Silurian System, removing almost the entire lower half, leaving what until now had been regarded as the Upper Silurian to represent the entire system. It also took a substantial bite out of Sedgwick's extended system, removing the entire Upper Cambrian.

Many in the geological establishment were sceptical.

MURCHISON VERSION		SEDGWICK VERSION		LAPWORTH VERSION	
UPPER SILURIAN	Ludlow	SILURIAN	Ludlow	SILURIAN	Ludlow
	Wenlock		Wenlock		Wenlock
LOWER SILURIAN	Bala/ Caradoc	UPPER CAMBRIAN	Bala/ Caradoc	ORDOV-ICIAN	Bala/ Caradoc
	Llandeilo		Llandeilo		Llandeilo
CAMBRIAN	Cambrian grits and shales	LOWER CAMBRIAN	Cambrian grits and shales	CAMBRIAN	Cambrian grits and shales

10.3 A comparison of the three versions of the stratigraphic table according to Murchison, Sedgwick and Lapworth, circa 1879.

Sedgwick and Murchison may recently have died, but their influence and authority – particularly Murchison's – were still very strong. It would take another twenty years before the Geological Survey finally accepted the Ordovician, and a further sixty years before it was accepted by the International Commission on Stratigraphy, the rock-dating arm of the International Union of Geological Sciences.[19] 'The official high priests of science are intolerant of contradiction,' ran a letter to a newspaper at the time, 'and woe betide the private worker who exposes it.'[20]

Today, a plaque near the door of the shepherd's cottage at Dobb's Linn reads: 'Birkhill Cottage where between 1872 and 1877 Charles Lapworth recognised the value of graptolites as a clue to the Geological Structure of these hills'. As one commentator would later observe, 'no other creature has done more to write history with their own physical remains'.[21]

But Lapworth was not finished. In 1881, at the age of thirty-nine, he was appointed the first professor of mineralogy and geology at Mason Science College, an institution dedicated

10.4 Charles Lapworth circa late 1880s. There seems to be a complete absence of pictures of the younger man.

to training people in science regardless of their background, and the forerunner of the University of Birmingham. The following year he decided to take his new zonal approach to that other area where there was still uncertainty about the nature and structure of the older rocks: the mountains and moors of northern Scotland.

The established view of the geology of the Highlands, set out by Murchison twenty years earlier, was almost universally accepted. Just as it had for the southern Highlands, it proposed a smooth succession of strata, in this case growing gradually younger as they passed from west to east, starting with ancient Primary rocks along the west coast and finishing with Devonian Old Red Sandstone on the east. In this way Murchison had been able to claim much of the central region, the bogs and moors of the Moine, or 'flow country', for Siluria.

But as James Nicol, Murchison's one-time collaborator, had pointed out, this ignored the fact that immediately below the highly metamorphosed schists and granites of the Moine lay a band of limestone and quartz, the Durness Limestone, which remained completely untouched by the heat associated with metamorphism. To resolve this geological implausibility, Nicol had talked of enormous movements in the Earth's crust, which brought great wedges of ancient metamorphic rock to the surface and left it sitting on top of unmetamorphosed quartz and limestones, leaving a huge fault line running roughly north–south through northern Scotland.

This meant there were two totally opposing versions of the geology of the region. As David Oldroyd, in his seminal book on the Highland Controversy has suggested, Lapworth 'would have regarded the north west Highlands as a grand opportunity to display again the benefits of his approach to fieldwork'.[22]

The Lapworth archive is housed in an eponymously named museum at the University of Birmingham, and is said to be one of the most complete archives devoted to a single geologist in

Britain. Shelf upon shelf of boxes contain fossils, notebooks, letters and hundreds of annotated maps, many of them still uncatalogued. There is no sign of any diaries, and as a result many of the particulars of Lapworth's fieldwork are missing, but we know that on 3 August 1882 he set out from Birmingham on a 600-mile journey to Scotland's far north coast. This was where the Scottish geologist Charles Peach had found what were considered to be Silurian fossils in beds of Durness Limestone, and we can presume that Lapworth saw it as a promising place to employ his zonal technique.

On the face of it the challenge was pretty simple: could he use his micro-fieldwork techniques to determine which account of the rocks was correct? But as Lapworth followed a series of low grassy banks where the Durness Limestone falls into the sea near the coastal village of Durness, he must have felt his heart sink: there was an almost complete absence of fossils, including graptolites, and those he did find were almost impossible to extract without damage. There was simply not enough data to apply the zonal techniques of the Southern Uplands.

He moved east, following the coast to 'the waters of the beautiful sea-fjord of Loch Eriboll' and its scattered crofting community, where he set up camp at a place known as 'Heilem Ferry', the site of a small inn that doubled as a booking hall and waiting room for the local ferry.[23] Today it's derelict, and a forlorn sign reminds visitors that Scotland's 'right to roam' laws do not include 'breaking into or entering' the property.

Navigating the shores of Loch Eriboll, stumbling over rocks slippery with bladderwrack, Lapworth could see chunks of crystalline gneiss, their minerals lined up in prominent black-and-white stripes, and lumps of reddish granite studded with pale crystals of mica. But any fossil record had been obliterated by the heat and pressure of metamorphism. He soon widened his area of search, trekking across miles of trackless and almost impassable grass and heather bog that stretched along the coast. Again the fossil record was sparse and obscure, and it soon become painfully clear to Lapworth he would have to fall back on more traditional methods. This was not going to be another triumph for graptolite zonation.

One area of bare rock soon caught his attention. Looking out of his room at the Heliam Ferry across the grey waters of the loch, he could see a small cliff on the opposite headland where changes in colour and texture indicated an obvious sequence of different strata. Lapworth climbed up through dense patches of heather and bracken, identifying and charting the distinct bands with his customary eye for detail. It was another slow and painstaking survey on hands and knees, stopping every few feet to hammer at the rocks and knock off the weathered surface to see what lay beneath. After several weeks of obsessively detailed work, he was able to draw a detailed cross-section through the cliff. It seemed to break naturally into three principal bands, something like this:

1. A top band of highly metamorphosed dull reddy/black Moine schist interwoven with layers of brilliant mica.
2. Several bands of pale chalky Durness Limestone and gritty sandstone.[24]
3. Grey Basal Quartz.

Here at last was what Lapworth was looking for: a Dobb's Linn of the north.

As before, he used it as a template, tracing out the same, but less prominent pattern in the rocks to the north and south of Eriboll and along the northern coast. He logged his findings in small, neat handwriting on six-inches-to-the-mile Ordnance

Survey maps, which he had cut up into convenient squares to make them easier to handle. And as he did so, two things rapidly became clear.

First, he came up against the same problem that had bothered Nicol: a bed of highly metamorphosed Moinian schists and gneiss, his top band, was sitting immediately on top of a bed of unmetamorphosed limestone and sandstones, which remained completely untouched by the heat and pressure of metamorphism. Second, the strata were heavily folded and faulted, a clear indication that there had been large movements in the Earth's crust. 'The folding, wrinkling, and inversion [of the rocks] in the Durness area is excessive,' Lapworth would later record. 'Almost every slab from the schistose and gneissose rocks exhibit the most extraordinary wrinkling and puckering.'[25] Murchison's 'simple' model of the region's rocks was, once again, coming unstuck.[26]

But there was something else, too. Protruding from the peat and heather Lapworth found boulders composed of dozens of fine lines or laminates. On close examination under a hand-lens, he could see that these were not bands of strata, but grains of the rock that had been crushed and 'smeared' by enormous pressure and heat. They seemed to suggest huge lateral movements in the Earth's crust that had split and ground the rocks together. 'Conceive a vast rolling & crushing mill of irresistible power,' he wrote to a friend, 'shale, limestone, quartzite, granite and the most intractable gneisses crumple up like putty in the terrible grip of this earth-engine, and are all finally flattened out into thin sheets of uniform lamination and texture.'[27] He called it Mylonite, from the Greek *mylon*, meaning 'mill', in recognition of the fact that the rock consisted of finely milled grains welded together by heat and pressure. To Lapworth this was powerful supporting evidence of Nicol's thesis that the rocks of north-west Scotland, far from representing a smooth succession of strata, had been subjected to massive faulting and movement: what he would later call 'gliding planes'.[28]

In the late summer of 1882 Lapworth returned to Birmingham to digest his findings. The following spring he read the first of what would become a series of talks to the Geologist's Association in London, entitled 'The Secrets of the Highlands'.[29] In his address he described the Moine schists as part of an ancient mountain region 'composed of rocks from very different geological ages', some of them pushed up from deep in the Earth's crust, which had been 'crushed and crumpled together by extensive lateral pressure, locally inverted, profoundly dislocated and partially metamorphosed'. He was envisaging, as Nicol had before him, a huge slab of ancient rock pushing up and over younger rocks in 'some appalling convulsion of nature – the vast area of which the wildest imagination fails to conceive', jumbling and tumbling them together along a massive fault line that ran through the region and created, in the process, beds of Mylonite.

'He possessed, to a remarkable degree, the faculty of thinking in three dimensions,' wrote a colleague later of Lapworth's talents.[30] 'Very few men' were able to study a piece of land and 'lay it before his hearers so that they seemed to see the rock framework under the landscape and vegetation'.[31]

Yet despite this powerful and evocative evidence, Lapworth – like Nicol before him – was cutting against the grain of orthodox thinking and a colleague warned him, 'You will be called upon for very strong evidence to substantiate your view ... the current of geological opinion ... tends to set in the opposite direction.'[32]

So it was that, a month later, as the days lengthened, Lapworth returned to northern Scotland to firm up his thoughts and examine the fault line in more detail. He was a talented

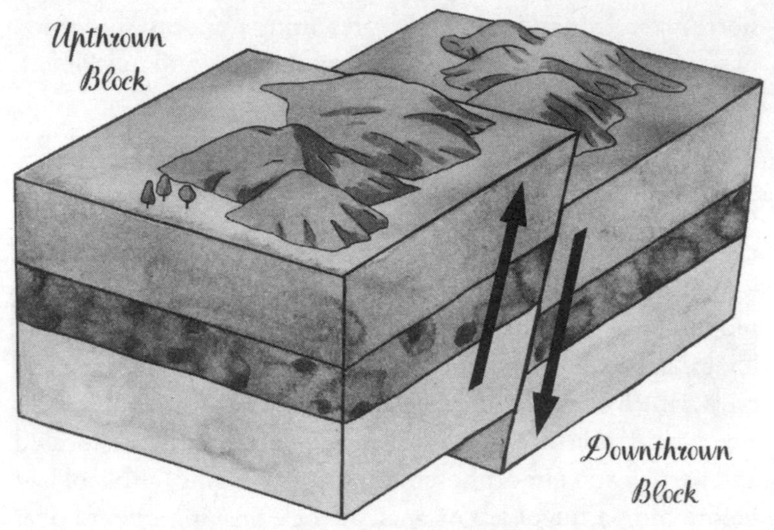

10.5 A classic 'overthrust' or 'reverse thrust' fault, such as that running through north-west Scotland.

draughtsman with an eye for the rise and fall of the land, and his carefully marked-up OS maps, supplemented by beautifully hand-drawn diagrams of the coast and countryside, show that he traced outcrops of Mylonite and their associated faulting for more than 100 miles across north-west Scotland, from the region of the Assynt and its strange red sandstone hills to the northern coast around Eriboll: the same fault line that Nicol had mapped twenty-six years earlier.

It was exhausting work across difficult and remote terrain, and the trek seems to have taken a terrible toll. Before he could complete his task, Lapworth was struck down by a mysterious illness. 'As he lay ill in the shepherd's hut below the Arnaboll Ridge [on Loch Eriboll] he used to fancy that the gneiss was still moving westward on the overthrust fault which he had discovered and mapped and that both he and the house would be

10.6 Lapworth's drawing of the north coast of Scotland.

crushed by its advancing mass,' wrote a friend.[33] The experience was so alarming that Lapworth dropped everything and fled, most uncharacteristically, back to his family in Birmingham.

We will never know exactly what happened. Was it the result of overwork or an infection? Lapworth was never physically robust. On the other hand, perhaps he had suffered some kind of psychological breakdown. His advocacy of Nicol's reverse-thrust fault meant he was taking on the geological establishment, including a man he had previously crossed swords with: Archibald Geikie, friend, colleague and later biographer of Roderick Murchison. Geikie had been the regional director of the Geological Survey in Scotland when Lapworth announced the concept of the Ordovician and forced the Survey to reappraise its understanding of the geology of the Southern Uplands

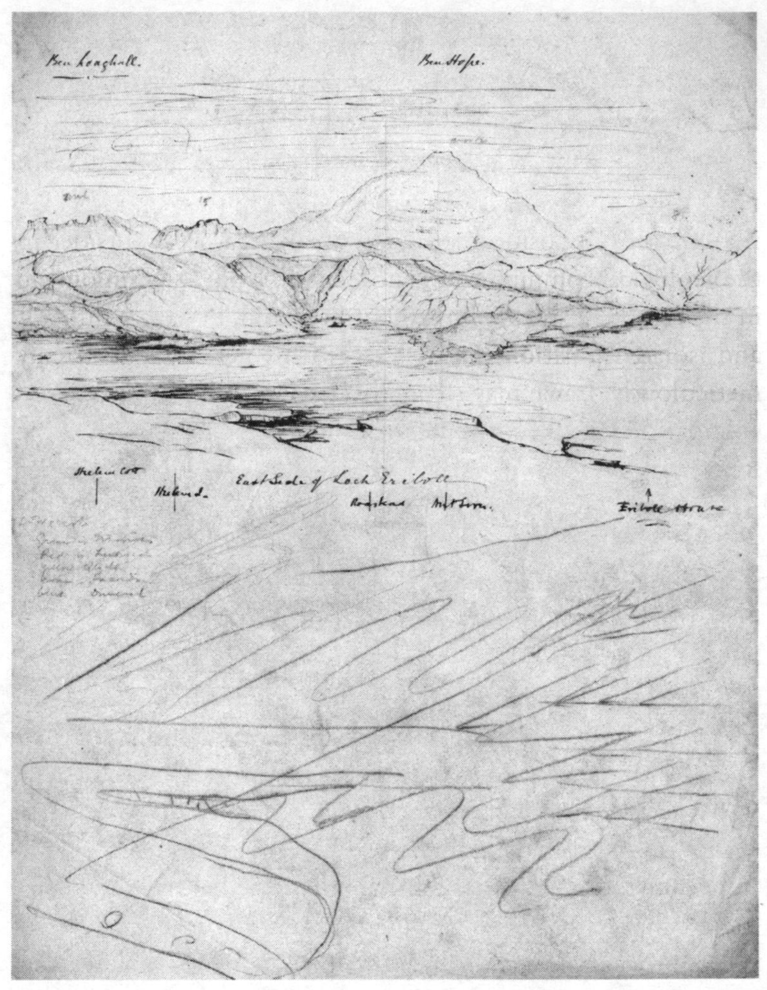

10.7 A Lapworth map from the summer of 1883, in which his usual meticulous cartography is uncharacteristically overlain by squiggles in purple crayon. Some observers have suggested it indicated that his mind was unbalanced.

four years earlier. Ever since, Geikie had continued to oppose the concept of the Ordovician at every opportunity. If Lapworth was right, Geikie – by now the national director of the Geological Survey – would be forced to a second major climbdown in almost as many years, at the hands of a man he still regarded as an amateur.[34] Could it be, as some have suggested, that Geikie's opposition left Lapworth, already an outsider in the geological community and now cut off from companionship in the remote fastness of northern Scotland, in such a difficult and isolated position that it affected his nerves? His usually meticulously drawn maps from this period contain curious and enigmatic child-like squiggles in purple crayon, which has led to speculation about the balance of his mind.

Yet there is little hard evidence that Lapworth ever worried much about Geikie's feelings. According to his contemporaries, Lapworth was a man who regarded the truth as 'a veritable religion',[35] and had so far shown little fear of the geological establishment. Indeed, only months earlier he had written a scathing review of Geikie's new geology textbook, arguing that the author was 'heavily handicapped ... by his amusing reluctance to admit that any British born subject who has not been, or is not, a member of H.M. Geological Survey, can possibly produce anything worthy of notice in geology'.[36] These are not the words of a man fearful of the social consequences of his actions. But whatever really happened in that remote shepherd's cottage, the implications of Lapworth's work were now abundantly clear: Murchison, and the Geological Survey, had missed a central geological feature of the Highlands. Geikie, with what must have been profound misgivings and now almost certainly under considerable pressure from his surveyors, reluctantly agreed to yet another new survey.

Twelve months later it confirmed what Lapworth – and

Nicol before him – had predicted: the established model of the geology of the Highlands was wrong. Instead of a smooth succession of rocks growing gradually younger as they passed from west to east, the region was the site of a huge 'overthrust' or 'thrust' fault, in which the rocks of the Moinian schists had been pushed up from deep in the Earth's crust and, moving westwards, ended up on top of the younger rocks of the Durness Limestone.

It was another extraordinary achievement, although the mysterious illness, combined with the Geological Survey's prompt return to the region, robbed Lapworth of some of the glory. Indeed, the Survey's official report of its findings published in the science journal *Nature* made no reference to Lapworth or his work. By all accounts, Lapworth was generous and open-hearted in victory and seems to have welcomed the new consensus: 'The old subject of dispute has utterly disappeared and there is no longer any reasonable excuse for dissension,' he would write. 'We have all been partly right and partly wrong. It is time for a hearty laugh all round, a time to shake hands and be friends.'[37]

Today this long belt of faulting running up through north-west Scotland is known as the Moine Thrust, a major fault line running 200 miles from the island of Skye up to the northern coast at Loch Eriboll. For most of its length the area of disturbance is around half a mile wide, but in places it spills out into a wider zone. I sat on a tomb in the ruined chapel at Balnakeil, a tiny cluster of buildings on the western fringes of Durness, watching waves break gently on the pale-yellow sand as oyster-catchers paddled in the shallow lagoons left by the receding sea. Hard as it was to imagine, the thrust zone passed pretty much directly below this peaceful scene. Here, when extinct arthropods such as trilobites and eurypterids ruled the late-Silurian

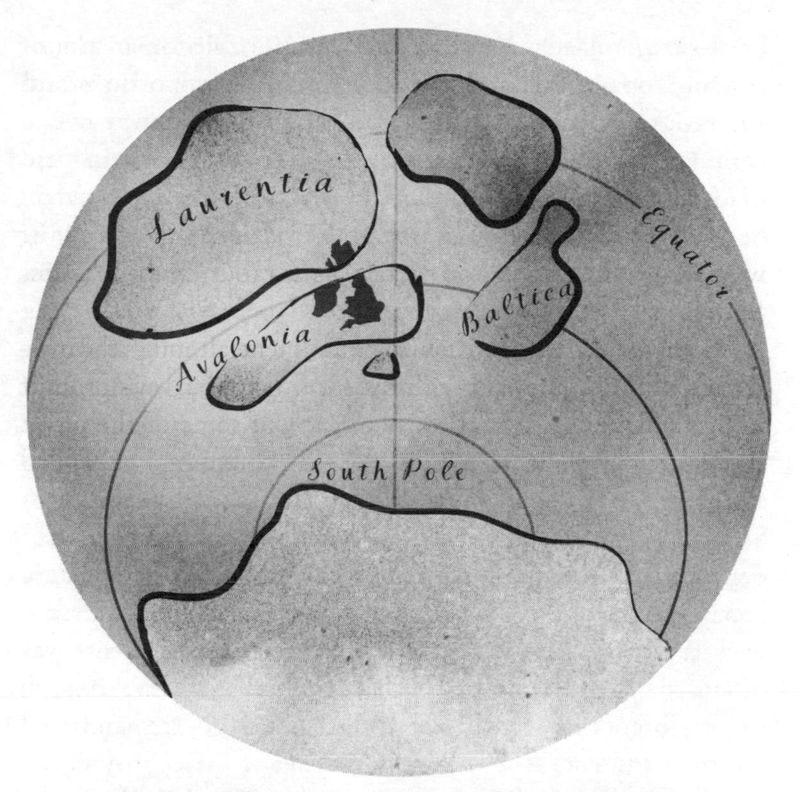

10.8 The fusing of Laurentia and Avalonia along the Iapetus Suture during the late-Silurian period about 430 million years ago.

world, two of the Earth's tectonic plates slowly and inexorably collided, melting, folding, grinding and faulting the rocks.

One of them was a relatively small tectonic plate called Laurentia, which contained parts of what we now call North America and the northern British Isles; the other was Avalonia, a tiny micro-continent containing, among other regions, southern Britain. As the plates collided, the coastal fringes of Laurentia were pushed down into the Earth's crust, while the westward-moving coast of Avalonia was pushed up and over them. As the

Earth's crust folded, what would become the Caledonian mountains of Scotland were pushed up to Himalayan proportions, and older rocks ended up on top of younger ones. In this way, over a period of 30 million years, northern and southern Britain were welded together. This is why there are marine fossils in the quartzite and limestone bands of northern Scotland that also occur in North America but are completely absent in southern Britain, while Lewisian Gneiss is closely related to rocks found in Canada and Greenland. Lapworth couldn't have known it at the time, but he had uncovered the birth of what we now call the British Isles.[38]

Back in Birmingham, Lapworth would eventually recover from his mysterious illness. He devoted the rest of his life to mapping the Ordovician Period in the Greywacke rocks of Shropshire and Wales, while teaching a steady stream of students interested in the now-mainstream subject of geology. He appears to have been quietly charismatic and, like Adam Sedgwick, a gifted lecturer. 'The sequence of geological events was disclosed in a series of pictures of a vividness so intense that the hearers forgot they were merely listening to words,' remembered a former student. 'They saw the sinking of the coastline, the deposition of the sediments, the sea vanquishing the land, the marvellous growth of corals – as if the speaker were a painter who had been present through the ages and could show the pictures he had painted and say "thus and no otherwise was it".'[39]

Lapworth died in 1920. He was buried, at his request, in Lodge Hill Cemetery in Birmingham, where he shares a simple and now-neglected grave with his wife and daughter in the shadow of a beech tree. His gravestone makes no mention of his remarkable achievements, but time has not forgotten him: he is recognised as a towering figure in British and world geology – a modest, warm and generous man who finally completed the puzzle that Sedgwick and Murchison had set out to solve.

Chapter Eleven

THE FINAL MYSTERY

The identification of the Cambrian, Ordovician and Silurian periods finally made it possible to catalogue and map the early history of the Earth. But one puzzle remained. Careful observers had noted that large areas of mainly Cambrian sandstone scattered across the hills and moors of north Wales – Sedgwick's 'crusty old rocks' – were composed of a strange and unusual assortment of different-sized grain particles, a mixture of sand, gravel and silt, which seemed to defy the normal processes of sedimentation.

Sandstone is usually laid down under water, according to a very simple principle: the greater the movement in the water, the larger the particles held in suspension. It follows that as the stream slows (think of a river flowing into the sea), first large and then ever-smaller particles drop to the bottom, creating a succession of beds composed of ever-finer grains of sand and silt. There is, in other words, a 'sorting' process, which ensures that similarly sized particles are laid down together and that gravel, sand and silt are not usually found mixed together in the same band of rock. Yet in parts of north Wales this mixing is exactly what appears to have happened.

Pick up a lump of dull red sandstone in the inhospitable upland moors of the Rhinog Hills in southern Snowdonia, a treeless wilderness of rocky peaks and heather crags, and it will almost certainly be composed of a blend of different-sized grains of pebbly sand, silt and rock fragments that have never been deposited or sorted by the usual process of sedimentation. For years, geologists wondered how this had happened. The final answer came, in totally unexpected circumstances and on the other side of the Atlantic Ocean, nearly half a century after Lapworth's death.

On 18 November 1929 a large earthquake, measuring 7.2 on the Richter scale, shook the sea floor of the Grand Banks fishing grounds off the coast of Newfoundland. It was felt thousands of miles to the south, in New York City. Forty-foot waves battered the Canadian shoreline, wiping out coastal settlements and killing tens of people. Within minutes, the transatlantic telegraph cables running between Britain and North America began to snap, the timing of each rupture exactly recorded by the break in telegraph traffic.

In the following months, when the timings of the breaks were scrutinised, a strange pattern emerged: the sequence of the ruptures bore no relation to the proximity of the different cables to the epicentre of the earthquake. Twelve telegraph lines in all were damaged, some running parallel with the coast in relatively shallow water, others further out in the ocean. Yet the record showed that the six cables closest to the shore, and therefore furthest from the quake, were the first to go, while the remaining six cables ruptured in a progressive sequence, one after another, as they got further out from the shore towards the epicentre of the tremors, with the last and deepest cable breaking thirteen hours and seventeen minutes later than the first.[1]

It seemed to fly in the face of logic. And why it should have happened remained a mystery for twenty-three years, until two American oceanographers revisited the data during a totally unrelated research project. In 1952 Bruce Heezen, a geologist, and Maurice Ewing, an expert in seismology, were mapping the ocean floor off the Atlantic coast of North America. Like many others before them, they became intrigued by a series of deep fissures, or submarine canyons, that ran down the coastal shelf like a series of valleys running off a mountainside. What could have caused them? The leading theory suggested that unstable banks of silt, brought down by rivers and deposited in shallow coastal waters, gradually grew in size until they 'toppled over', sending avalanches of sand and mud crashing down the continental shelf, scouring out canyons as they went.[2] For obvious reasons, though, nobody had ever witnessed such a phenomenon and it remained, at best, a guess.

Heezen and Ewing, casting round for sources of data of underwater disturbances, now revisited the records of the Grand Banks earthquake. It was a revelation. Looking at the puzzling sequence of breaks in the transatlantic cables, they realised the pattern was almost certainly not a direct result of the earthquake itself, but was exactly what might be expected if tremors from the quake had disturbed unstable banks of silt that had accumulated in the shallow waters of the shoreline, sending an underwater avalanche of silt and mud crashing down the coastal shelf. The leading edge of the avalanche would have ruptured the six cables closest to the shore and then, as it gathered speed and continued down the shelf, accumulating an ever-denser load of sand and silt as it went, it would have gouged out underwater canyons and ruptured the six remaining cables further out to sea, one after another. This was not only the first plausible explanation for the pattern of breaks in the cables, but the first tangible evidence, albeit indirect, that underwater avalanches might be real.[3]

Today there are still no direct observations of these so-called 'turbidity currents', or underwater avalanches, but repeated

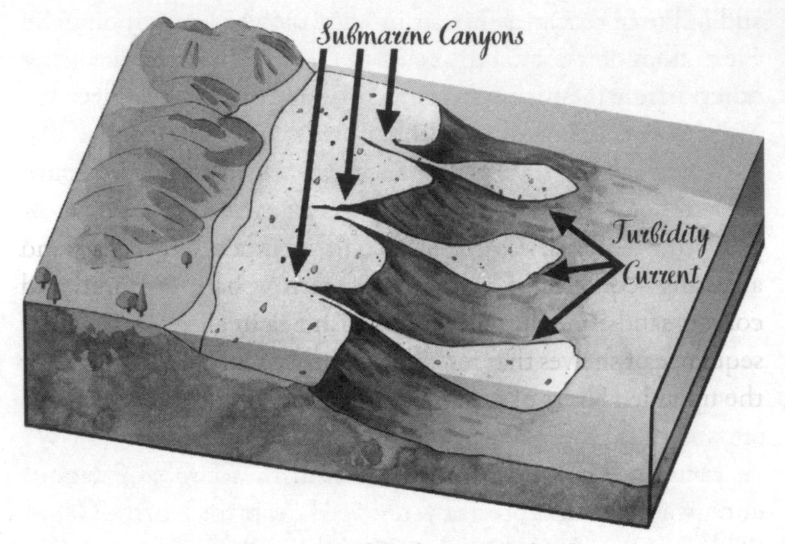

11.1 An illustration of the formation of undersea canyons and the nature of turbidity currents.

laboratory simulations using a blend of silt and sand have shown that the assortment of sediments deposited by such experiments – a sandstone jumble of sand, gravel, grit and silt, sometimes deposited in rough irregular bands – is almost identical to core samples taken from sites like the Grand Banks.[4] It's also remarkably similar to the composition of the puzzling rocks found in places like the Rhinog Hills.

This observation has led geologists to conclude that some 400 million years ago, when these unusual rocks were laid down during the Cambrian, Ordovician and Silurian periods, this region of Wales must have been part of a shallow coastal shelf. Fast-flowing rivers running off a nearby land mass to the east dumped huge quantities of sand and silt in the sea, generating turbidity currents that raced down the coastal shelf. As they settled on the ocean floor, they deposited the mix of sand, silt and mud that we now associate with the Welsh Greywacke. Here at last was a final explanation for the puzzling rocks that had been tormenting geologists for more than a century. The Welsh Greywacke would seem to have yielded up its last secret.

One of the most remarkable places to see the results of such an avalanche is north of the small Welsh seaside town of Aberystwyth, where a series of coastal cliffs composed of a crazy and arresting sequence of hundreds of narrow bands of mud and coarser sandstone slump into the Irish Sea in a jagged, irregular sequence of shelves that resemble tightly packed beds of slate, or the upended layers of a slice of millefeuille.

As I finished writing this book I returned to Wales and the moss-encrusted limestone pillars at Gelli-grin on the western slopes of the Berwyns. In the deep quiet of the Welsh hills, with only a buzzard calling overhead to puncture the silence, I thought again about the eminent Victorian geologists who had fought and struggled over these unimaginably ancient rocks.

Roderick Murchison, in particular, has cast a long shadow over the world of geology. As director-general of the British Geological Survey until his death in 1871, he was able to use his considerable influence to ensure that much of his geological nomenclature would live on. As a result, the names of small towns and geological features in the Welsh Marches can be found in rocks across the world, in particular the Ludlow, Wenlock and Llandovery series of the Silurian System and, in some versions of the stratigraphic table, in the Caradoc epoch of the Ordovician System.[5] And in case we were in any danger of forgetting his name, there are at least thirteen geographical features across the globe called after Roderick Murchison, many of them legacies of Britain's colonial past. They include four Mount Murchisons: one in New Zealand, where there is also a town named after him, one in Tasmania, one in Canada and one in Antarctica; a Murchison Range of mountains in Greenland, New Zealand and South Africa, where there is also a mining town called Murchison; the Murchison Falls on the upper reaches of the White Nile in Uganda; the Murchison River in Western Australia; and the tiny Murchison Island off the Pacific coast of northern British Columbia. In addition there's a Murchison Crater on the near side of the moon; a Murchison memorial in the Siberian city of Perm, recalling his naming of the Permian Period in 1842; and a blue plaque on a handsome town house in the Pennine market town of Barnard Castle reminding visitors that Murchison lived in the town between 1818 and 1822, soon after he had left the army.

In the years immediately following his death in 1871, however, his reputation among his contemporaries was rather less glorious. A review of Archibald Geikie's two-volume biography of Murchison in the *Edinburgh Review* argued that, 'in our opinion Murchison was not the greatest but the most successful of the geologists of his time ... his scientific successes were not those of a man of commanding genius, but they were the results of indomitable industry and business-like method'.[6] The surveyor (and later academic) Joseph Beete Jukes was even harsher: 'Sir Roderick was never anything more than a man who had an eye for the country but was a mere pettifogger in original observation and knows no more than a wooden spoon.'[7]

By comparison, Adam Sedgwick's nomenclature has vanished from the world's geological maps, yet he was widely regarded by his contemporaries as a much more original and significant scientist. The *Edinburgh Review* lamented that 'the man of genius had been, as is generally the case, outstripped in the race by the practical man of method. It was undoubtedly very hard for Sedgwick to feel that his labour of years was, to some extent, unrequited, while his friend had achieved with comparative ease a great geological victory.' Beete Jukes, as ever, went further, describing Sedgwick as 'a grand original observer and logical reasoner', touched by 'the divine gift of genius'.

Sedgwick's name lives on in the Sedgwick Museum of Earth Sciences in Cambridge, and in the University's geological club, the Sedgwick Society, said to be the oldest student-run geological association in the world. Soon after his death the citizens of Sedgwick's home village of Dent in the Yorkshire Pennines erected a large granite water fountain in his memory, which still sits in the marketplace. And in 1985 the Yorkshire Dales National Park laid out a Sedgwick Geological Trail several miles to the north of Dent, to mark the bicentenary of his birth.

Fittingly, the trail follows the valley of the shallow rocky River Clough, where the fast-flowing stream has exposed the underlying strata and it is possible to trace the lines of the so-called Dent Fault, a fault line running through the underlying rocks where, some 350 million years ago during the early Carboniferous Period the earth's tectonic plates shifted, continents collided and the ancient Silurian rocks of the Lake District were pushed up so that today they sit alongside the younger Carboniferous Limestones of the Yorkshire Dales. It was a fault line that Sedgwick had done much to expose, even if he never fully understood its significance.

The least well remembered and celebrated of my three protagonists is the brilliant Charles Lapworth. The sole geological feature named after him is a long-vanished glacial lake on the River Severn in northern Shropshire, called Lake Lapworth. Today all that remains of it are a series of glacial gravel deposits in the Shropshire plain. His principal memorial, besides the Lapworth chair of geology at the University of Birmingham, is the University's Lapworth Museum of Geology where, in a large, windowless store room lined with shelves, his notebooks, maps and letters still await a biographer to do justice to his quite remarkable achievements.

Over a period of fifty years, each one of these men would claim the perplexing calcareous ashes of the Gelli-grin for their own: a microcosm of a wider debate that would bring out the best, and the worst, in the newly emerging science of geology and would gradually advance our understanding of the last 500 million years, when life as we know it started to emerge on Earth.

As I tramped through deep beds of dead leaves, brown and curled in the summer sun, I picked up a tiny fragment of this very ordinary-looking rock: dull grey, without any outstanding features. I keep it as a memento of an extraordinary story, 'one of the most lively and innovative periods in the history of the earth sciences,' as one historian put it.[8] In testing their wits against the riddles of the Greywacke, Sedgwick, Murchison and Lapworth all taught us to read the past in the present. They rewrote the story of our planet, revealing it to be grander and stranger than we had ever supposed. And they brought the science of geology out of its infancy, leading it into the wild, trackless country that is the true place of adventure and discovery.

GLOSSARY

Anticline/Syncline

An anticline is where rock strata are pushed up to create a dome. A syncline is where they are folded down to form a depression or trough.

Gneiss, schist

Metamorphic rocks that look superficially similar where extreme heat and pressure have caused the minerals to line up in distinctive layers or bands. Schist is coarser grained than gneiss.

Greywacke

Originally a Victorian catch-all term used to describe the unmapped ancient rocks found across Europe including, in Britain, large areas of Wales, Scotland, the Lake District and the West Country. Also known as Transition Rocks. Today the term refers to sandstone rocks composed of a jumble of different grain sizes.

Gypsum

A crystalline rock formed when water rich in calcium and sulphates evaporates, a similar process to that which creates salt pans when sea water evaporates.

Igneous

Formed when magma or volcanic lava cools and crystallises, either on the Earth's surface or inside the crust. Examples include granite, basalt and pumice.

Magnesian Limestone

Rock in which the calcium carbonate in limestone has been enriched or replaced by magnesium carbonate in a process that is still unclear.

Metamorphic, metamorphism

Rocks transformed by heat and/or pressure into a different kind of rock, often from non-crystalline into crystalline. The slower the rate of cooling the bigger the crystals. Examples include quartzite, gneiss and schist.

Mylonite

A metamorphic rock in which the grains have been crushed and stretched by intense movements in the Earth's crust, giving them a streaked or smudged appearance.

Porphyry

A Victorian term for crystalline or igneous rocks. Now used to refer to rocks where large crystals sit in a bed of smaller ones.

Quartzite

A hard quartz-rich sandstone transformed by heat and pressure into a crystalline rock.

Sedimentary

Rocks laid down as sediments, usually underwater. Common ones include chalk, limestone, sandstone and mudstone.

Shale

A fine-grained sedimentary mixture of mud and limestone.

Strike/Dip

Terms used to describe the angle and orientation of a line of strata. The strike is the compass orientation of the strata: whether it runs north, east, south or west. The dip is the angle the strata sits at relative to the horizontal.

NOTES

Introduction

1. William Fitton and Roderick Murchison, 'The Silurian System', *Edinburgh Review*, April 1841.
2. See, for instance, the biblical calculations of James Ussher, the seventeenth-century Primate of All Ireland and archbishop of Armagh, in J. A. Carr, *The Life and Times of James Ussher: Archbishop of Armagh*, London, Wells, Gardner, Darton & Co., 1895.
3. Jacquetta Hawkes, *A Land*, Boston, Beacon Press, 1951, p. 51.

1 The Erratic Professor Sedgwick

1. Hon. John Byng, *The Torrington Diaries 1781–1794*, 4 vols, edited by C. Bruyn Andrews, New York, Barnes & Noble, and London, Methuen, 1970.
2. See: ukfossils.co.uk/2012/07/12/wrens-nest/
3. Information from the Lapworth Museum, University of Birmingham.
4. John Willis Clark and Thomas McKenny Hughes, *The Life and Letters of the Reverend Adam Sedgwick*, 2 vols, Cambridge, Cambridge University Press, 1890, Vol. 1, p. 378.
5. Darwin had been tutored at Cambridge by the naturalist and philosopher William Whewell, a friend and colleague of Sedgwick. Whewell encouraged Sedgwick to help his young student. Darwin's interest in geology may have been encouraged by his grandfather, the polymath Erasmus Darwin, who had surveyed the rocks at Ironbridge and discovered tar in them.
6. Colin Humphrey, 'Cambria versus Siluria: A Dispute over the Emerging Geology of Wales', Mid Wales Geology Club. Humphrey writes that 'Darwin's elder sister, Susan, was bowled over by the attractive and fascinating Sedgwick and Darwin's biographer writes that her brother joshed her about whether there was to be an announcement about a forthcoming Mrs Sedgwick.'

7. This concept of the Tertiary should not be confused with the contemporary term, which includes consolidated rocks.
8. Roderick Murchison, *Siluria*, London, John Murray, 1854, p. 6.
9. Archibald Geikie, *Life of Sir Roderick Murchison*, London, John Murray, 1875, Vol. 1, p. 180.
10. Quotes taken from William Fitton and Roderick Murchison, 'The Silurian System', *Edinburgh Review*, April 1841. See also Archibald Geikie, *Founders of Geology*, London, Macmillan and Co., 1897.
11. William Conybeare and William Phillips, *Outlines of the Geology of England and Wales*, London, William Phillips, 1822.
12. Henry De la Beche, *Geological Manual*, London, Charles Knight, first published in 1831.
13. Charles Lyell, *Principles of Geology*, 3 vols, London, John Murray, 1831–3.
14. Today we know these early geologists were only scraping the surface of the Earth's history. Modern dating techniques show that the category 'Primary rocks' lumped together around four *billion* years of the planet's early history, and that the categories of Tertiary, Secondary and Transition rocks were merely describing the most recent 500 million years. But the Primary rocks, mangled and distorted by the enormous passage of time, their tiny fossils crushed and obscure, were simply beyond the comprehension of early geologists. The Greywacke was regarded as about as far back in the history of the Earth as it was possible to go, and it was felt that a proper understanding of it would not only reveal the history of the rocks, but would also show when – and perhaps even how – life on Earth had begun.
15. See David R. Oldroyd, *Earth, Water, Ice and Fire: Two hundred years of geological research in the Lake District*, London, Geological Society Memoir, No. 25, 2002.
16. In the Sedgwick Museum in Cambridge there is a fine collection of Sedgwick's hammers, ranging from heavy-duty sledgehammers with long handles and blunt heads, which would have required two hands to manipulate them, to smaller, lighter single-handed hammers with short handles and sharp pick-like heads. There is also a bronze statue of Adam Sedgwick standing guard at the junction of the two wings of the museum, where he is holding his trusty geological hammer in one hand.
17. For more on Darwin in Wales, see Michael Roberts, 'Charles Darwin's geological fieldwork in Wales', *Endeavour*, Vol. 25 (1), 2001; see also Clark and Hughes, *Life and Letters of the Reverend Adam Sedgwick,* Vol. 1, p. 381.
18. Geikie, *Life of Sir Roderick Murchison*, Vol. 1, p. 178.
19. Clark and Hughes, *The Life and Letters of the Reverend Adam Sedgwick*, Vol. 1,

p. 378. Letters, Sedgwick to Murchison, 13 September 1831, Geological Society of London [GSL]: M/S11/51 8.

20. We now know this gap represented some 40 million years of the Earth's history.
21. Geikie, *Life of Sir Roderick Murchison*, Vol. 1, p. 178.
22. Letters, Sedgwick to Murchison, 13 September 1831, GSL: M/S11/51.
23. Ibid.
24. There is some uncertainty about their mode of transport, but Sedgwick is known to have preferred to ride wherever possible. See Oldroyd, *Earth, Water, Ice and Fire*.
25. Nicholas Crane, *The Making of the British Landscape*, London, Weidenfeld & Nicolson, 2016, p. 431.
26. Clark and Hughes, *The Life and Letters of the Reverend Adam Sedgwick*, Vol. 1, p. 380.
27. Ibid., Vol. 1, p. 378.
28. In 1830 there were only two professional geologists in Britain: Sedgwick in Cambridge and William Buckland at Oxford University. A third part-time academic post in geology was created at King's College London in 1831.
29. Quoted in Martin J. S. Rudwick, *The New Science of Geology. Studies in the Earth Sciences in the Age of Revolution. A collection of articles and essays*, Aldershot, Ashgate, 2004, Ch. XI, p. 244.
30. The Sedgwick Club, http://sedgwickclub.soc.srcf.net/adamsedgwick.php
31. Ibid.
32. Clark and Hughes, *The Life and Letters of the Reverend Adam Sedgwick*, Vol. 1, pp. 160–1.
33. Ibid., Vol. 1, p. 182.
34. The professorship was vacant because the previous incumbent, the Rev. John Hailstone, proposed 'to take to himself a wife'.
35. Clark and Hughes, *The Life and Letters of the Reverend Adam Sedgwick*, Vol. 1, pp. 153–4.
36. In the words of one commentator, unlike a contemporary university professor 'who is expected to direct a department of scholars engaged in teaching and research, a Cambridge professor, at the turn of the nineteenth century, worked entirely on his own and his teaching duties, if any, were not related in any way to the examination syllabus of the university'. Colin Speakman, *Adam Sedgwick – Geologist and Dalesman*, Broad Oak Press, 1982, p. 54.
37. Ibid., p. 54.
38. Geikie, *Life of Sir Roderick Murchison*, Vol. 1, p. 138.
39. Clark and Hughes, *The Life and Letters of the Reverend Adam Sedgwick*, Vol. 2, p. 489.

40. Ibid., Vol. 2, p. 486.
41. Ibid., Vol. 1, pp. 308–9.
42. Ibid., Vol. 2, p. 486.
43. Ibid., Vol. 2, p. 486.
44. Geikie, *Life of Sir Roderick Murchison*, Vol. 1, pp. 205–6.
45. Ibid., Vol. 2, p. 350.
46. See Roberts, 'Charles Darwin's geological fieldwork in Wales'.
47. Esther Moir, *The Discovery of Britain: The English Tourists 1540–1840*, London, Routledge & Kegan Paul, 1964, p. 135.
48. The Rhinogs represent the most complete exposure of Cambrian rock in Britain. See for example Ian Vince, *The Lie of the Land*, London, Pan Books, 2010, Ch. 14.
49. K. M. Lyell (ed.), *Life, Letters and Journals of Sir Charles Lyell, Bart*, London, John Murray, 1881, Vol. 1, p. 367.
50. Quoted in Frances Kilvert, *Kilvert's Diary 1870–1879: Selections from the Diary of The Rev. Francis Kilvert*, edited by William Plomer, London, Vintage Classics, 2013, p. 125.
51. Clark and Hughes, *The Life and Letters of the Reverend Adam Sedgwick*, Vol. 1, p. 377.
52. At Llanllyfni near Caernarfon. Letters, Sedgwick to Murchison, 13 September 1831, GSL.
53. Patricia Pierce, *Jurassic Mary: Mary Anning and the Primeval Monsters*, Stroud, The History Press, 2006, p. 41.
54. Geikie, *Life of Sir Roderick Murchison*, Vol. 1, p. 191.
55. Sedgwick told a colleague in 1831 that he could roughly plot a baseline where the 'slates and grits' of north Wales gave way to the younger rocks of Snowdon. We now know such a baseline is an oversimplification. We don't know what kind of topographical maps he was using at the time to map the region. See David Oldroyd, *Earth, Water, Ice and Fire*.
56. Letter to Murchison, 20 October 1831, in Clark and Hughes, *The Life and Letters of the Reverend Adam Sedgwick*, Vol. 1, p. 382.

2 The Ambitious Mr Murchison

1. Geikie, *Life of Sir Roderick Murchison*, Vol. 1, p. 180.
2. Buckland made the identification at around the same time as the Lyme Regis fossil-collector Mary Anning, but in the masculine world of British geology he claimed, and won, the credit. See Pierce, *Jurassic Mary*, p. 104.
3. Geikie, *Life of Sir Roderick Murchison*, Vol. 1, pp. 125–6.
4. The forerunner of the military training college at Sandhurst.

5. The half a million figure comes from Clark and Hughes, *The Life and Letters of the Reverend Adam Sedgwick*, Vol. 2, p. 238.
6. Ibid., Vol. 1, p. 137.
7. Geikie, *Life of Sir Roderick Murchison*, Vol. 2, p. 323.
8. Today Murchison's house in the small Teesdale market town of Barnard Castle bears a plaque with an inscription that includes the passage: 'Formerly the residence of Sir Roderick Murchison, twice president of the Royal Geographical Society. He died in 1871 aged 79 a great geologist and explorer.'
9. James A. Secord, 'King of Siluria: Roderick Murchison and the Imperial Theme in nineteenth century British Geology', *Victorian Studies*, Vol. 25. See also Anthony Brook, 'Aspects of Murchison', West Sussex Geological Society, Occasional Papers No. 2, December 2001.
10. See John Stafford, *Scientist of Empire: Sir Roderick Murchison, scientific exploration and Victorian imperialism*, Cambridge, Cambridge University Press, 1989, for a discussion of Murchison's financial circumstances at this time.
11. John J. Morton, *King of Siluria: How Roderick Murchison changed the face of geology*, Horsham, Brocken Spectre Publishing, 2004, p. 33.
12. The Royal Institution of Great Britain is an organisation devoted to scientific education and research, based in London.
13. See Speakman, *Adam Sedgwick – Geologist and Dalesman*, p. 67.
14. James A. Secord, *Controversy in Victorian Geology: The Cambrian–Silurian Dispute*, Princeton, NJ, Princeton University Press, 1986, p. 45.
15. See, for example, Brook, 'Aspects of Murchison'; Stafford, *Scientist of Empire*, Ch. 8; and Rudwick, *The New Science of Geology*, Ch. XI, p. 255. Other ex soldiers included the mapmaker George Greenough; the founder of the British Geological Survey, Henry De la Beche; the retired major-general Joseph Portlock; the palaeontologist William Lonsdale; and the explorer Sir Henry James. See also Secord, *Controversy in Victorian Geology*, p. 61.
16. De la Beche, *Geological Manual*, p. 600.
17. Secord, *Controversy in Victorian Geology*, p. 44.
18. James David Forbes, an Edinburgh physicist, quoted in Secord, 'King of Siluria', p. 41.
19. In the early days of geology many 'local' and 'amateur' geologists remained obscure and unpublished because they lacked the funds to publish and propagate their findings. See, for instance, Hugh Torrens, 'Scientific ancestry and historiography of the Silurian System', *Journal of the Geological Society*, Vol. 147, 1990.
20. The Laki eruption in southern Iceland in June 1783 blasted lava and ash into the atmosphere for eight months, covering Europe in a dense cloud of dust and fog

that altered the climate for the remainder of the summer and following winter. The clergyman and naturalist Gilbert White watched events unfold from Selborne, a small village in Hampshire. The summer that year, he recorded, 'was an amazing and portentous one, and full of horrible phaenomena; for besides the alarming meteors and tremendous thunder-storms that affrighted and distressed the different counties of this kingdom, the peculiar haze, or smokey fog, that prevailed for many weeks ... [meant] ... the sun, at noon, looked as blank as a clouded moon, and shed a rust-coloured ferruginous light on the ground' (Gilbert White, *The Natural History of Selborne*, London, Bell & Daldy, 1870, p. 33). The US politician Benjamin Franklin, at that time a diplomat in Paris, recorded similar clouds of dust blocking out the sun: 'There existed a constant fog over all Europe, and a great part of North America. This fog was of a permanent nature; it was dry, and the rays of the sun ... were indeed rendered so faint in passing through it, that when collected in the focus of a burning glass they would scarce kindle brown paper' (Benjamin Franklin, 'Meteorological imaginations and conjectures', *Memoirs of the Literary and Philosophical Society of Manchester*, 1st series, Vol. 2, pp. 359–60).

21. Robert Macfarlane, *Mountains of the Mind*, London, Granta Books, 2004, p. 53.
22. Hutton's model was among the more comprehensive theories of the Earth's origins, but several European thinkers had come up with similar ideas, in particular the French polymath Georges-Louis Leclerc, Comte de Buffon, a naturalist and cosmologist, who reached a very similar set of conclusions in the late 1700s.
23. The alternative concept, with its biblical undertones, of the Earth being the product of a one-off catastrophic event, usually thought of as a flood, became known as 'Catastrophism'. It was an attractive idea to an age that was still heavily influenced by the Church.
24. They visited in particular the volcanic hills of the Auvergne and the Massif Central in France, where the power of volcanoes to alter and shape the landscape was all too apparent.
25. Professor James Secord, in his unpublished account of Murchison's 1831 trip to the Wye Valley, suggests there may also have been a third reason for the trip to Wales. He writes that Murchison's wife's mother was in poor health, and the couple may have wanted to stay in England to be close at hand in case anything happened, rather than take one of their usual trips to the Continent. James A. Secord, 'A Romance of the Field: Roderick Murchison's Geological Discovery of 1831', unpublished.
26. Kilvert, *Kilvert's Diary*, p. 62.
27. Geikie, *Life of Sir Roderick Murchison*, Vol. 1, p. 182.

28. Ibid., Vol. 1, p. 183.
29. Arthur Murchison, a possible distant relative of Murchison, has written about this. 'A point I attempt to make in [my book] *War Before Science* is that RIM's role models from his earliest years into his 20s was military men who were born and grew up in the 18th century, especially his uncle, General Sir Alexander Mackenzie, as well as many of the other military men in his family, regiment and brigade. Beside his uncle, Gen. Mackenzie, General John Marchant, founder and Superintendent of the Royal Military College in Marlow, his regimental colonel, Colonel R. Burne, General Sir Arthur Wellesley (later the Duke of Wellington) was at the top of his admiration list; he was in Wellington's presence three times and wrote about those occasions with obvious relish years later. Thus we have a lad with a military upbringing and education surrounded by military officers, all of whom were proper red coated officers and gentlemen, of the previous century. Those and many other military officers served as young RIM's role-models to greater or lesser degrees. An obituary of Murchison in the Edinburgh University Library supports this view. "There was always something of the soldier about him."' Personal correspondence.
30. See Secord, *Controversy in Victorian Geology*, p. 61.
31. Secord, 'King of Siluria', p. 421.
32. Daniel Defoe, 'From Chester to Holyhead', published as an appendix to Daniel Defoe's *From London to Land's End*, 1722.
33. Today there is almost nothing left of Harpton Court except a gatehouse and drive. Frankland Lewis would later become a member of the Geological Society. J. B. Sinclair and R. W. D. Fenn, 'Geology and the Border Squires', *Proceedings of the Radnor Society*, Vol. 69, 1999, pp. 143–72.
34. Edward Evans was one of several local experts that Murchison would employ during his field trips.
35. Sinclair and Fenn, 'Geology and the Border Squires', p. 154.
36. These volcanic hills are part of what's known today as the Builth Inlier, an area of older rock surrounded by younger. Murchison will explore it in some detail over the following years, logging some 30–40 different marine fossil species. It is particularly famous for trilobites.
37. Brenda Colloms, *Victorian Country Parsons*, Lincoln, NE, University of Nebraska Press, 1977, p. 116.
38. Lewis's presidential address to the Woolhope Naturalists' Field Club in 1854, quoted in John Fuller and Hugh Torrens, 'Murchison in the Welsh Marches: A History of Geology Group field excursion led by John Fuller, May 8th–10th, 1998', Shropshire Geological Society.

39. We'll see later that the reality is more complicated, but in the early 1800s the concept of a unique 'fossil fingerprint' was enormously important.
40. Daniel Defoe, 'From Chester to Holyhead'.
41. Ian Vince writes that 'the gentleman scientists of Ludlow were among the first in the world to set up a natural history society in 1833, an offshoot of which is the Ludlow Museum, which opened 50 years before London's Natural History Museum'. Vince, *The Lie of the Land*, p. 211.
42. We now know this is not true, but it was good enough at the time.
43. *Caledonian Mercury*, 6 October 1831. See also Geikie, *Life of Sir Roderick Murchison*, Vol. 1, p. 189.
44. Ibid., Vol. 1, pp. 184, 189.
45. Ibid., Vol. 1, p. 183.
46. Ibid., Vol. 1, p. 184.
47. Murchison is known to have used a number of local stonemasons and quarry men as guides, in particular around Llandeilo in the Towy Valley and in the volcanic hills around Builth Wells.
48. Geikie, *Life of Sir Roderick Murchison*, Vol. 1, p. 192.
49. Ibid., Vol. 1, p. 181.
50. Samuel Woodward, quoted in Speakman, *Adam Sedgwick – Geologist and Dalesman*, p. 70.
51. Letter from Lewis to Murchison, 16 July 1854, quoted in Sinclair and Fenn, 'Geology and the Border Squires', pp. 143–72.
52. Murchison's original plan had been to visit East Anglia, where Lyell was planning an investigation into the Tertiary muds and clays of the region.
53. Secord, *Controversy in Victorian Geology*, p. 55.
54. The location is also known as Cavansham Ferry.
55. Hugh Torrens has recorded that Sedgwick would later vehemently maintain Murchison had had no particular plan to explore the Greywacke in 1831, explaining that Murchison had written to him that his main object was 'to see the pranks the igneous rocks were playing'. Torrens, 'Scientific ancestry and historiography', pp. 657–62. Secord would also observe, in *Controversy in Victorian Geology*, p. 56: 'given the importance of the 1831 tour for both his career and for the history of science, it is not surprising that Murchison should view it as the outcome of a deliberate search'.
56. I am grateful to James Secord for this information, which appears in his unpublished 'A Romance of the Field'.
57. Personal correspondence.
58. See Fuller and Torrens, 'Murchison in the Welsh Marches'; and Duncan Hawley,

'The first true Silurian: An evaluation of the site of Murchison's discovery of the Silurian', *Proceedings of the Geologists' Association*, Vol. 108, pp. 131–40.

59. Geikie, *Life of Sir Roderick Murchison*, Vol. 1, p. 138.
60. Lyell to Scope, 9 November 1830, in Lyell (ed.), *Life, Letters and Journals of Sir Charles Lyell*, Vol. 1, pp. 309–11 at p. 310. Quoted in Secord, *Controversy in Victorian Geology*, p. 46.
61. See, for instance, Humphrey, 'Cambria versus Siluria'.

3 Trekking through Wales

1. Admiral George Brydges Rodney: https://en.wikipedia.org/wiki/George_Brydges_Rodney,_1st_Baron_Rodney
2. Clark and Hughes, *The Life and Letters of the Reverend Adam Sedgwick*, Vol. 1, p. 392.
3. Byng, *The Torrington Diaries*.
4. Clark and Hughes, *The Life and Letters of the Reverend Adam Sedgwick*, Vol. 1, p. 392.
5. He called it 'Gaderferwyn', but I assume it's what we know as Cadair Berwyn.
6. I'm grateful to Dr Charles Bendall for the following information. Today the name Bala Limestone has been dropped, and these rocks are known as the Cymerig Limestone Member of the Gelli-grin Calcareous Ash Formation. They are in the Upper Ordovician – Caradoc Series Woolstonian Stage. The limestone is interbedded with volcanic ashes, shales/slates and gritty sandstones.
7. In fact we now know Sedgwick was wrong about the two strata being identical, a mistake that would lead to problems in the future.
8. By 1966 a total of eighty-eight different trilobite and brachiopod species had been found in the two limestones.
9. Sedgwick would write to Murchison on 21 July 1832, 'Through the whole ascent, and nearly to the base on the W. side of the chain, the dip is about W by N ... (while) ... working gradually to NW in the prolongation of the chain towards Corwen. On the western side of the chain an anticlinal line strikes through the region NNE and WSW, in consequence of which some bands of black shelly limestone I found at the top of the Berwyns are brought out again with an opposite dip viz ESE.' Clark and Hughes, *The Life and Letters of the Reverend Adam Sedgwick*, Vol. 1, p. 392.
10. See Clark and Hughes's description of the Bala Limestone. Ibid., Vol. 2, p. 515.
11. *Letters and Extracts from the Addresses and Occasional Writings of J. Beete Jukes, M.A.*, edited by his sister, C. A. Browne, London, Chapman and Hall, 1871.
12. Ibid.
13. Andrew Ramsay, *Memoirs of the Geological Survey*, London, Longmans, 1866,

p. 87. The copy of this book held in University College London library was presented to the university by Murchison himself in 1866.

14. Colin Humphrey, Personal correspondence. Interestingly, James Secord has written: 'Despite its straightforward appearance on geological maps (and in my own account up to this point) the Bala Limestone could not be observed in the field as a continuous bed with a single obvious orientation. Rather like most inland strata in Wales it was visible only as a set of temporary quarries and discrete outcrops assembled by the geologist according to a set of tacit rules and grouped under a single name. For a full understanding one had to be taken over the ground by someone already familiar with the exposures.' Secord, *Controversy in Victorian Geology*, p. 154.
15. For more on Sedgwick's fieldwork, see Oldroyd, *Earth, Water, Ice and Fire*.
16. The small booklet *Introduction to Foxhunting* describes an 'eye for the country' as the ability 'without apparent thought, always to go the right way when riding to hounds'. D. W. E. Brock, *Introduction to Foxhunting*, London, Seeley, Service & Co., Beaufort Library, 1954, Vol. III.
17. Geologists at the time assumed they were identical. We now know the Aymestry Limestone is slightly older than the Ludlow.
18. Macfarlane, *Mountains of the Mind*, p. 43.
19. Arthur Truman, *Geology and Scenery in England and Wales*, London, Penguin, 1971, p. 18.
20. Hugh Torrens has pointed out that by the early 1830s the rocks of southern Shropshire were relatively well explored and the early industrialisation of the Shropshire coalfields had encouraged geological activity. Pioneer geologists included Thomas Du Gard of Shrewsbury (1777–1840), a founding member of the Geological Society, and Robert Townson (1762–1827), who self-published *A Sketch of the Mineralogy of Shropshire* in 1799. See Torrens, 'Scientific ancestry and historiography', pp. 657–62. Other early pioneers were Romney and Robson Wright, two brothers working for the Ministry of Defence as mapmakers.
21. For encouragement to publish his pamphlet, see Clark and Hughes, *The Life and Letters of the Reverend Adam Sedgwick*, Vol. 1, pp. 324–5. For the self-published pamphlet, see Arthur Aikin, 'Proposals for a mineralogical survey of the county of Salop, and of some adjacent districts', 1810. Aikin's pamphlet would provide Murchison with invaluable background information, but he would receive scant acknowledgement in the years to come. The *Edinburgh Review* was particularly struck by the similarity between Aikin's cross-sections and Murchison's, remarking, 'Mr Aikin's notes appear to us to coincide with Mr Murchison's descriptions'. See Hugh Torrens, 'Geological Pioneers in

the Marches', *Proceedings of the Shropshire Geological Society*, Vol. 13, 2008, pp. 65–76.

22. The map was compiled by the brothers Romney and Robson Wright. I'm indebted to Mike Rosenbaum of the Shropshire Geological Society for this information. See 'On the Secondary Formations in the neighbourhood of Ludlow' by J. R. Wright, *Proceedings of the Geological Society*, London, 1832.
23. Geikie, *Life of Sir Roderick Murchison*, Vol. 1, p. 206.
24. Kilvert, *Kilvert's Diary*, p. 50.
25. Ibid.
26. Geikie, *Life of Sir Roderick Murchison*, Vol. 1, p. 217.
27. Quoted in Moir, *The Discovery of Britain*, p. 127.
28. Entry for 11 June 1928, in Lyell (ed.), *Life, Letters and Journals of Sir Charles Lyell*, Vol. 1, p. 189.
29. Murchison's journals are held in the Geological Society archive in London, MN 63, p. 66.
30. Letter to Murchison, 23 July 1832, in Clark and Hughes, *The Life and Letters of the Reverend Adam Sedgwick*, Vol. 1, p. 393.
31. Letter to William Ainyer, student friend from St Bees, ibid., Vol. 1, p. 395.
32. Letter to Murchison, 23 July 1832, ibid., Vol. 1, p. 393.
33. Ibid., Vol. 1, p. 281.
34. Ibid., Vol. 1, p. 395.
35. See Byng, *The Torrington Diaries.*
36. Clark and Hughes, *The Life and Letters of the Reverend Adam Sedgwick*, Vol. 1, p. 395.
37. Byng, *The Torrington Diaries.*
38. Walter Scott, *Red Gauntlet*, London, Macmillan & Co., 1925.
39. Clark and Hughes, *The Life and Letters of the Reverend Adam Sedgwick*, Vol. 1, p. 316.
40. Ibid., Vol. 2, p. 498.
41. Sedgwick's letters to friends would contain long and minute accounts of his attempts to deal with gout, including avoiding 'strong drink', which at other times he seems to have taken to excess, and 'dephlogistication' (putting out the fire), a medical procedure designed to reduce inflammation.
42. *Letters and Extracts from the Addresses and Occasional Writings of J. Beete Jukes.*
43. Clark and Hughes, *The Life and Letters of the Reverend Adam Sedgwick*, Vol. 1, pp. 392–3.
44. Ibid., Vol. 1, p. 207.
45. Secord, *Controversy in Victorian Geology*, p. 80. Sedgwick to Murchison, GSL: M/S11/73.

46. Clark and Hughes, *The Life and Letters of the Reverend Adam Sedgwick*, Vol. 1, p. 409.
47. Geikie, *Life of Sir Roderick Murchison,* Vol. 1, p. 219.
48. According to his field notes, he traced the Trilobite Flags from Llandeilo to Llandovery, to Llandrindod Wells, to Rorrington and Hope (east of Montgomery), to Longnor (south of Shewsbury) to Horderly (east of Oswestry) and presumably up to somewhere like Glyn Ceriog. See Murchison's notebooks for 1833, GSL: M/N64–8.

4 A New History: the Cambrian and Silurian

1. Quoted in Secord, *Controversy in Victorian Geology*, p. 90.
2. Murchison to Whewell in July 1834, Clark and Hughes, *The Life and Letters of the Reverend Adam Sedgwick*, Vol. 1, p. 430.
3. Kilvert, *Kilvert's Diary.*
4. Speakman, *Adam Sedgwick – Geologist and Dalesman*, p. 71.
5. Sedgwick extolled Élie de Beaumont's 'parallelism' in his capacity as president of the London Geological Society and incorporated it into his lectures in Cambridge. See Secord, *Controversy in Victorian Geology*, p. 65.
6. Geikie, *Life of Sir Roderick Murchison*, Vol. 1, p. 222.
7. Ibid., Vol. 1, p. 228.
8. Clark and Hughes, *The Life and Letters of the Reverend Adam Sedgwick*, Vol. 1, p. 433.
9. Ibid., Vol. 1, p. 439.
10. Ibid., Vol. 1, p. 440.
11. Ibid., Vol. 1, p. 234.
12. Sedgwick to Murchison, 28 October 1827, quoted in Clark and Hughes, *The Life and Letters of the Reverend Adam Sedgwick*, Vol. 1, p. 306.
13. James Woodforde, *The Diary of a Country Parson 1758–1802*, Norwich, Canterbury Press, 1999, p. 68.
14. Adrian Desmond and James Moore, *Darwin: The Life of a Tormented Evolutionist*, London, Norton & Company, 1991, p. 54. I'm grateful to Arthur Murchison for drawing my attention to this.
15. Geikie, *Life of Sir Roderick Murchison*, Vol. 1, p. 223.
16. Charles Lyell writing in 1832, quoted in Clark and Hughes, *The Life and Letters of the Reverend Adam Sedgwick*, Vol. 1, p. 385.
17. For more on Sedgwick and women, see ibid., Vol. 1, pp. 453–4.
18. Ibid., Vol. 1, p. 513.
19. Ibid., Vol. 2, p. 25.

20. The exact process under which slatey cleavage occurs is still unclear, although there has been some speculation that it is caused by mild metamorphism.
21. Clark and Hughes, *The Life and Letters of the Reverend Adam Sedgwick*, Vol. 1, p. 386.
22. Ibid., Vol. 1, p. 387.
23. Humphrey, 'Cambria versus Siluria'.
24. Cited in Morton, *King of Siluria*, p. 68. The *Philosophical Magazine* started publication in 1798 at a time when philosophy meant 'natural philosophy', which embraced all aspects of science.
25. Murchison to Whewell, 20 November 1834, quoted in Secord, *Controversy in Victorian Geology*, p. 100.
26. Martin J. S. Rudwick, *The Great Devonian Controversy*, Chicago, Chicago University Press, 1985, p. 133.
27. The paper was called 'On the Silurian and Cambrian systems, exhibiting the order in which the older Sedimentary Strata succeed each other in England and Wales'. Sedgwick had also considered calling his system the 'Snowdonian', but, he told Murchison, he regarded it as a 'beastly modern word of the Saxon tourists'.
28. Rudwick, *The Great Devonian Controversy*, p. 101.
29. Taken from Secord, *Controversy in Victorian Geology*.
30. Clark and Hughes, *The Life and Letters of the Reverend Adam Sedgwick*, Vol. 1, p. 445.
31. Edward Forbes was an active member of the Geological Society. Quoted in Secord, *Controversy in Victorian Geology*, p. 198.
32. Geikie, *Life of Sir Roderick Murchison*, Vol. 1, p. 218.

5 The Devonian Controversy

1. De la Beche was said to have had an inheritance worth £3,000 p.a. (more than £300,000 p.a. today) from Jamaican sugar. His father, Thomas Beech, had changed the family name in 1790, almost certainly to enhance the family's social status.
2. Keith Shaw, 'Sir Henry Thomas De La Beche in Lyme Regis', Lyme Regis Museum. See also Paul J. McCarthy, 'Henry De la Beche: Observations on an Observer', Friends of the National Museum of Wales, Cardiff, 1977.
3. Pierce, *Jurassic Mary*.
4. The manuscript of 'The Lymiad' was published for the first time by the Lyme Regis Museum in 2011. It was said to have been written by a local lawyer called John Doble Burridge. The poem is the copyright of the Lyme Regis Museum.
5. The Slavery Abolition Act abolished slavery in the British Empire in 1833.

6. De la Beche's *A Geological Manual* was published in 1831 and quickly ran into three English editions, followed by French, German and American versions.
7. De la Beche, *A Report on the Geology of Cornwall, Devon and West Somerset*, HMSO, 1839, p. 515.
8. John Lindley had co-authored, with John Claudius Loudon, an *Encyclopedia of Plants* published in 1829. It was a massive undertaking and covered nearly 15,000 species of flowering plants and ferns.
9. Rudwick, *The Great Devonian Controversy*, p. 99.
10. Murchison would almost certainly have known about the Culm in Devon, and the 1833 edition of De la Beche's *Geological Manual* had reported similar deposits of anthracite in the Greywacke in France and Germany.
11. Greenough to De la Beche, December 1834, quoted in John C. Thackray (ed.), *To See the Fellows Fight*, Norwich, British Society for the History of Science, 1999, p. 59.
12. Rudwick, *The Great Devonian Controversy*, p. 99.
13. Thackray, *To See the Fellows Fight*, p. 60.
14. Ibid., p. 108.
15. A report from the English Heritage Extensive Urban Survey, 'An Archaeological Assessment of Newport' by Clare Gathercole, 2002, suggested that Minehead was short on 'gracious lodgings'.
16. Murchison notebook 22, p. 23, quoted in Rudwick, *The Great Devonian Controversy*, p. 151.
17. Murchison recorded in his journal for 1836 that although he found organic remains, he was unable to identify them with anything Silurian. Today these rocks are known to contain the fossilised remains not just of crinoids, but of bivalves such as cockles and mussels, brachiopods and small slugs and snails or gastropods.
18. Murchison reported the location as being near the small hamlet of Pilton, now a suburb of Barnstaple. He would describe it as 'a very good representative of the Llandeilo Flags', which, he said, formed 'the base of the true Silurian System'.
19. Moir, *The Discovery of Britain*, p. 11.
20. Geikie, *Life of Sir Roderick Murchison*, Vol. 1, p. 251.
21. Quoted in Rudwick, *The Great Devonian Controversy*, pp. 154–5.
22. The fossils included round pinkish corals, crinoids, brachiopods, squids and octopus-like cephalopods, and a distinctive chamber-shelled ammonite called clymenia. The quarry at South Petherwin is now largely filled in and difficult to explore.
23. An enthusiastic local fossil-collector called Richard Hennah had unearthed

fossils of the marine brachiopod spirifera, which is common in younger Carboniferous Limestone.

24. Sedgwick to a friend, 20 July 1836, quoted in Clark and Hughes, *The Life and Letters of the Reverend Adam Sedgwick*, Vol. 1, p. 459.
25. The recipe appears in Sedgwick's field notebook no. 21. 1831.
26. Letter to William Whewell, 12 October 1836, in Clark and Hughes, *The Life and Letters of the Reverend Adam Sedgwick*, Vol. 1, p. 463.
27. The quarry was probably what is known today as Lower Dunston Quarry. The dates of the visit are a little uncertain. There is nothing in Sedgwick's notes, and Murchison would later complain that his notes on the Devon trip had been lost.
28. Geikie, *Life of Sir Roderick Murchison*, Vol. 1, p. 251.
29. Murchison's presentation was entitled 'A classification of the Olde Slate Rocks of Devonshire and on the true position of the Culm deposits of the central portion of the country'.
30. Quoted in Rudwick, *The Great Devonian Controversy*, p. 166.
31. In January 1837 a local geologist called Major William Harding had found Carboniferous, Culm-type fossils, including ferns and horsetails, near the small Devonshire village of Marwood, just north of Barnstaple. It was in an area Sedgwick and Murchison had designated as much older rock, probably 'Upper Cambrian'. Two months later the Rev. David Williams, an indefatigable fossil-hunter and local vicar in the Mendip village of Bleadon, found similar fossilised ferns and land-based plants in the surrounding hill, an area that Sedgwick and Murchison had identified as Silurian.
32. The fossils included trilobites, brachiopods (lamp shells) and crinoids or sea lilies. When they were examined by John Phillips at London University, he declared them 'strangely analogous' to well-known Carboniferous species. Phillips had just completed a monograph on the Carboniferous Limestone in Yorkshire and was arguably Britain's greatest authority on the subject. Sedgwick also knew Austin and described him as a 'clever, good, independent workman'.
33. Quoted in Rudwick, *The Great Devonian Controversy*, pp. 195–6.
34. Ibid., p. 189.
35. Geikie, *Life of Sir Roderick Murchison*, Vol. 1, p. 253.
36. Ibid., Vol. 1, p. 254.
37. See Stafford, *Scientist of Empire*, p. 19.
38. Geikie, *Life of Sir Roderick Murchison*, Vol. 2, p. 158.
39. It's a member of the Spiriferida family of brachiopods.
40. Secord, *Controversy in Victorian Geology*, p. 120.
41. Rudwick, *The Great Devonian Controversy*, p. 262.

42. Lyell, *Principles of Geology*, first published in 1830 and then regularly updated and reprinted, with the final edition printed posthumously in 1875.
43. William Whewell's February 1838 address to the Geological Society.
44. Letter to a clerical friend, quoted in Humphrey, 'Cambria versus Siluria'.
45. Clark and Hughes, *The Life and Letters of the Reverend Adam Sedgwick*, Vol. 1, p. 465.
46. Letter to a clerical friend, quoted in Humphrey, 'Cambria versus Siluria'.
47. Clark and Hughes, *The Life and Letters of the Reverend Adam Sedgwick*, Vol. 1, p. 466.
48. Speakman, *Adam Sedgwick – Geologist and Dalesman*, p. 51.
49. Quoted in Torrens, 'Scientific ancestry and historiography', pp. 657–62.
50. Both these examples are taken from Humphrey, 'Cambria versus Siluria'.
51. Henry De la Beche, *A Report on the Geology of the West Country of Cornwall, Devon and West Somerset*, London, Longman, Orme, Brown, Green and Longmans, 1839.
52. Rudwick, *The Great Devonian Controversy*, p. 265.
53. Murchison to Sedgwick, February 1839, quoted ibid., p. 269.
54. We now know, of course, that the early history of the Earth stretches back much further, into the so-called Precambrian. But in the nineteenth century this period was unknown.
55. Geikie, *Life of Sir Roderick Murchison*, Vol. 1, p. 269.
56. The rocks also contained the remains of two curious armoured fish, Coccosteus and Pterichthys.
57. The fossil record in these rocks was largely composed of brachiopods and trilobites.

6 Stumbling on the Permian in Russia

1. Michael Collie and John Diemer, *Murchison's Wanderings in Russia*, British Geological Survey, 2004, p. 23.
2. P. S. Pallas, *Bemerkungen auf einer Reise in die Südlichen Statthalterschaften des Russischen Reichs*, Leipzig, 1799–1801; see also *Travels through the Southern Provinces of the Russian Empire*, St Petersburg, 1771–6
3. Between May and November 1829 von Humboldt spent more than six months travelling from the Neva Valley in western Russia to the Yenisei River in central Siberia. See, for example, Geological Society of London, *Special Publications*, 2007, 287 (1), pp. 161–75.
4. Murchison kept a detailed journal of the trip, which he hoped to publish as a popular travelogue, but it was an uncomfortable mix of personal anecdote and raw field data. *Murchison's Wanderings in Russia* was finally published in 2004

by the British Geological Survey, edited and annotated by Michael Collie and John Diemer. Much of this chapter, including these details, is drawn from this invaluable source.

5. Geikie, *Life of Sir Roderick Murchison*, Vol. 1, p. 318.
6. Ibid., Vol. 1, p. 296.
7. Ibid., Vol. 1, p. 297.
8. Collie and Diemer, *Murchison's Wanderings in Russia*, p. 40.
9. Ibid., p. 50.
10. Geikie, *Life of Sir Roderick Murchison*, Vol. 1, p. 301.
11. Collie and Diemer, *Murchison's Wanderings in Russia*, p. 50.
12. Ibid., p. 66.
13. Ibid., p. 47.
14. Ibid., p. 46.
15. Geikie, *Life of Sir Roderick Murchison*, Vol. 1, p. 296.
16. Ibid., Vol. 1, p. 300.
17. Collie and Diemer, *Murchison's Wanderings in Russia*, p. 70.
18. Ibid,. p. 83.
19. It would later be identified as a bivalve called avicula.
20. Collie and Diemer, *Murchison's Wanderings in Russia*, p. 93.
21. Murchison's journal tells us this was the town of Vologda.
22. Collie and Diemer, *Murchison's Wanderings in Russia*, p. 83.
23. Ibid., p. 100.
24. Ibid., p. 108.
25. During the Soviet years it was renamed Gorky and became a centre of military production and research, and one of the Soviet Union's closed cities.
26. Collie and Diemer, *Murchison's Wanderings in Russia*, p. 83.
27. Ibid., pp. 114–15.
28. The fossils included nautiloids such as Orthoceratites, brachiopods like Productida and sea urchins such as Echinus that Murchison had last seen at Archangel.
29. Geikie, *Life of Sir Roderick Murchison*, Vol. 1, p. 302.
30. Collie and Diemer, *Murchison's Wanderings in Russia*, p. 126.
31. Ibid., p. 129.
32. The reign of Tsar Nicholas I was marked by repression, economic stagnation, poor administrative policies, a corrupt bureaucracy and frequent wars of imperial expansion.
33. Collie and Diemer, *Murchison's Wanderings in Russia*, p. 126.
34. Geikie, *Life of Sir Roderick Murchison*, Vol. 1, p. 319.
35. Ibid., Vol. 1, p. 327.

36. Collie and Diemer, *Murchison's Wanderings in Russia*, p. 188.
37. Geikie, *Life of Sir Roderick Murchison*, Vol. 1, p. 327.
38. Collie and Diemer, *Murchison's Wanderings in Russia*, p. 188. This is the well-known Vyazniki section, named after the small town of Vyazniki on the River Klyazma, where colourful bands of gypsum, shelly limestone, sandstone and marl were clearly visible. Murchison later examined specimens of the rocks under a powerful hand-lens and found 'a profusion of microscopic crustaceans resembling Cytherinae, associated with a small flattened bivalve shell, having the general form of Cyclas'. See Michael J. Benton et al., 'Murchison's first sighting of the Permian, at Vyazniki in 1841', *Proceedings of the Geologists' Association*, Vol. 121, Issue 3, 2010, pp. 313–18.
39. Collie and Diemer, *Murchison's Wanderings in Russia*, pp. 190–91.
40. Geikie, *Life of Sir Roderick Murchison*, Vol. 1, p. 327.
41. Collie and Diemer, *Murchison's Wanderings in Russia*, p. 251.
42. Ibid., p. 204.
43. Ibid., p. 210.
44. Ibid., p. 213.
45. Ibid., p. 247.
46. Ibid., pp. 224–5.
47. Ibid., p. 229.
48. Geikie, *Life of Sir Roderick Murchison*, Vol. 1, p. 329.
49. Collie and Diemer, *Murchison's Wanderings in Russia*, p. 249.
50. Geikie, *Life of Sir Roderick Murchison*, Vol. 1, pp. 330–31.
51. Murchison called it the River Uralsk.
52. Geikie, *Life of Sir Roderick Murchison*, Vol. 1, p. 341.
53. Ibid., p. 341. This is a reference to the British Association for the Advancement of Science.
54. Collie and Diemer, *Murchison's Wanderings in Russia*, p. 347.
55. Ibid., pp. 347–8.
56. Or the town of Saratoft, as Murchison called it.
57. Collie and Diemer, *Murchison's Wanderings in Russia*, p. 347.
58. Ibid., p. 359.
59. Ibid., p. 385.
60. Ibid., p. 377.
61. Ibid., p. 391.
62. Ibid., p. 410.
63. Ibid., p. 412.
64. Geikie, *Life of Sir Roderick Murchison*, Vol. 1, p. 354.
65. Murchison and de Verneuil's paper provided lists of marine fossils from the

Kazan, Perm and Orenburg. Many were regarded as intermediate in character between Carboniferous and Triassic life forms. Three years later, in Murchison's great book *The Geology of Russia*, published in 1845, he described the discovery of the key fossils at Vyazniki on the River Klyazma. 'In the light red, variegated marls towards the middle of the cliff, and in a ravine to the north of the highroad, we detected a profusion of microscopic crustaceans resembling cytherinae, associated with a small flattened bivalve shell, having the general form of *Cyclas*.' See Benton et al., 'Murchison's first sighting of the Permian', pp.313–18.

66. The German naturalist and explorer Alexander von Humboldt made a series of seminal journeys through regions of South America in the early 1800s. The British in India systematically surveyed the Himalayas in the mid-1800s. And in the 1850s a British Royal Geographical Society expedition found the source of the Nile.
67. See, for instance, Archibald Geikie, 'Murchison's labours among the older rocks stood indeed to geology in a relation not unlike that of his friend Livingstone's work in Africa.' Geikie, *Life of Sir Roderick Murchison*, Vol. 2, p. 346.
68. Geikie quoted in Secord, *Controversy in Victorian Geology*, p. 122.
69. Ibid., p. 120.

7 Siluria v. Cambria

1. See Chapter 1 (p. 19) for more on John Woodward and his interest in geology.
2. In 1835 Sedgwick had received nearly £2,000 from the proceeds of a slave-owner's will for whom he had been co-executor. He had used the money to buy specimens for the collection.
3. Clark and Hughes, *The Life and Letters of the Reverend Adam Sedgwick*, Vol. 2, p. 349.
4. The Cockerell Building was named after its chief architect, Charles Cockerell.
5. Clark and Hughes, *The Life and Letters of the Reverend Adam Sedgwick*, Vol. 2, p. 350.
6. The Geological Society occupied part of London's Somerset House, together with the Navy, the Stamp Office and several other learned societies.
7. Secord, *Controversy in Victorian Geology*, p. 113.
8. Sedgwick called his paper a 'Supplement' to his 1838 'Synopsis', and it was formally known as the 'Supplement to a Synopsis of the English Series of Stratified Rocks inferior to the Old Red Sandstone'.
9. Secord, *Controversy in Victorian Geology*, p. 129.
10. Ibid., p. 131.
11. Ibid, p. 129.

12. Geikie, *Life of Sir Roderick Murchison*, Vol. 2, p. 348.
13. Quoted in Secord, *Controversy in Victorian Geology*, p. 123. For a slightly different view of this, see Michael Collie and John Diemer, *Murchison in Moray: A Geologist on Home Ground*, Philadelphia, American Philosophical Society, 1995.
14. Secord, 'King of Siluria', p. 417.
15. Pierce, *Jurassic Mary*, pp. 132–79.
16. Geikie, *Life of Sir Roderick Murchison*, Vol. 1, p. 356.
17. Quoted in Secord, *Controversy in Victorian Geology*, p. 123.
18. Clark and Hughes, *The Life and Letters of the Reverend Adam Sedgwick*, Vol. 2, p. 494.
19. Ibid., p. 496.
20. Letter to Kate Malcolm, 6 January 1844. Ibid., pp. 65–6.
21. Ibid., p. 9.
22. Kate Malcolm was the daughter of Sir John Malcolm, a retired army officer and Member of Parliament who frequently hosted Sedgwick at the family home in Berkshire. Sedgwick befriended the young girl – initially, it is said, by carrying her on his back – and they continued to correspond until he died.
23. Letter to Kate Malcolm, October 1847. Clark and Hughes, *The Life and Letters of the Reverend Adam Sedgwick*, Vol. 2, p. 129.
24. Ibid., Vol. 2, p. 34.
25. Letter to Murchison, January 1843. Ibid., Vol. 2, p. 54.
26. Geikie, *Life of Sir Roderick Murchison*, Vol. 1, p. 365.
27. Ibid., Vol. 1, p. 363.
28. Secord, *Controversy in Victorian Geology*, p. 122.
29. Ibid., p. 127.
30. Ibid., pp. 127–9.
31. Ibid., p. 288.
32. Ibid., p. 248.
33. Clark and Hughes, *The Life and Letters of the Reverend Adam Sedgwick*, Vol. 2, p. 48.
34. Much of the scanty record of this trip is contained in notebook no. 37, which covers the two weeks from 10 to 26 September 1842. The account of the merry days spent with Salter comes from a second, similar trip the two men took in 1843 and appears in an unpublished account left by Sedgwick. It is quoted in Speakman, *Adam Sedgwick – Geologist and Dalesman*, p. 79.
35. For a sense of how irregular and erratic these limestone bands were, see Secord, *Controversy in Victorian Geology*, p. 154, on the Bala Limestone.
36. GSL: M/S11/209a & b.

37. During the same summer of 1842 Murchison also made a quick trip to north Wales with his Russian travelling companion Count von Keyserling. They worked their way north from Malvern, 'lingering at the rocks and country-houses which had been Murchison's favorite haunts ten years before', and crossed the boundary between the Silurian and Cambrian rocks that he and Sedgwick had agreed upon in 1834. This time, however, Murchison felt unable to draw 'any satisfactory line between the two tracts ...' and on that basis spread the colours of the lower Silurian to the western coast of Wales. The following year the Society for the Diffusion of Useful Knowledge published his first map of the region, which showed an extended Siluria.
38. *Quarterly Review*, March 1846, Vol. 77, pp. 348–80.
39. Secord, *Controversy in Victorian Geology*, p. 173.
40. The review appeared in the Scottish periodical *The North British Review*, a magazine with a liberal editorial policy, published by the Free Church of Scotland.
41. Secord, 'King of Siluria', p. 425.
42. Geikie, *Life of Sir Roderick Murchison*, Vol. 2, p. 42.
43. Sedgwick to Murchison, 7 February 1846, GSL: S/M 1846.
44. William Conybeare to Murchison, 21 December 1841, GSL, Murchison mss.
45. See Pierce, *Jurassic Mary*, p. 162.
46. Ibid., pp. 162–3.
47. Ibid., p. 157. It appears in Anning's so-called Common Place Book, an early form of autograph book in which she invited friends to write contributions.
48. See Sinclair and Fenn, 'Geology and the Border Squires', p. 151.
49. The cavern was known as the Dark Cavern and, once it had been worked out, it was equipped with gas lighting by Lord Derby and used for concerts, balls and lectures.
50. *Illustrated London News*, 22 September 1849, p. 201.
51. Midland Union of Natural History Societies, December 1891, p. 268, quoted in Oldroyd, *Earth, Water, Ice and Fire*.
52. Quoted in *Following in the footsteps of Murchison: A field excursion for the History of Geology Group 18th–20th July 2014*.
53. Geikie, *Life of Sir Roderick Murchison*, Vol. 2, p. 4.
54. Gideon Mantell writing to his friend Benjamin Silliman in June 1834, quoted in Torrens, 'Scientific ancestry and historiography', *Journal of the Geological Society*, Vol. 147, 1990.
55. Geikie, *Life of Sir Roderick Murchison*, Vol. 1, p. 403.
56. See, for instance, Rudwick, *The Great Devonian Controversy*, Ch. 7.
57. Secord, *Controversy in Victorian Geology*, p. 133.

58. Ibid., p. 147.
59. Ibid.
60. For nearly ten years so-called 'amateur' geologists – part-time geologists who were not regarded as part of the full-time elite – had been intensively exploring the Bala and Caradoc strata and had concluded they were probably the same. But they had not been listened to. Indeed one of them, a man called Daniel Sharpe, who made repeated field trips to the Berwyn Hills, had been heavily frowned upon by members of the Geological Society who regarded his fieldwork as intruding on what was considered Sedgwick's territory. It was seen as ungentlemanly conduct, even 'scientific trespass' and, in the words of one historian of the period, 'threatened a carefully balanced system of paternalistic exchanges between provincial amateurs and geological society experts'. Puns on Sharpe's name – 'the Sharpe fellow', 'a Sharpe move', 'too Sharpe for us' – abounded in the correspondence of the time. Quoted in Secord, *Controversy in Victorian Geology*, p. 163.
61. Clark and Hughes, *The Life and Letters of the Reverend Adam Sedgwick*, Vol. 2, p. 215.
62. Secord, *Controversy in Victorian Geology*, p. 215.
63. The paper was called 'On the Classification and Nomenclature of the Lower Palaeozoic Rocks of England and Wales', 25 February 1852.
64. Clark and Hughes, *The Life and Letters of the Reverend Adam Sedgwick*, Vol. 2, p. 215.
65. Secord, *Controversy in Victorian Geology*, p. 218.
66. Ibid., p. 220.
67. Gideon Mantel to Benjamin Silliman, quoted in Secord, *Controversy in Victorian Geology*, p. 215.
68. Clark and Hughes, *The Life and Letters of the Reverend Adam Sedgwick*, Vol. 2, p. 215.
69. Geikie, *Life of Sir Roderick Murchison*, Vol. 2, p. 140.
70. Secord, *Controversy in Victorian Geology*, p. 184.
71. Ibid., p. 223.
72. Clark and Hughes, *The Life and Letters of the Reverend Adam Sedgwick*, Vol. 2, p. 218.
73. Secord, *Controversy in Victorian Geology*, p. 231.
74. Murchison to Sedgwick, 22 November 1852, in Geikie, *Life of Sir Roderick Murchison*, Vol. 2, p. 142.
75. Ibid., Vol. 1, p. 359.
76. Secord, *Controversy in Victorian Geology*, p. 226.

77. Letter to Sedgwick, quoted in Clark and Hughes, *The Life and Letters of the Reverend Adam Sedgwick*, Vol. 2, p. 218.
78. Sedgwick to Beete Jukes in 1856, quoted in Secord, *Controversy in Victorian Geology*, p. 226.
79. Ibid., p. 170.
80. Ibid., p. 239.
81. Ibid., p. 239.
82. Quoted in Rudwick, *The Great Devonian Controversy*.

8 From Collaboration to Rivalry

1. Letter from Leonard Horner, a former secretary and president of the Geological Society, to Murchison, quoted in Secord, *Controversy in Victorian Geology*, p. 232.
2. Geikie, *Life of Sir Roderick Murchison*, Vol. 2, p. 141.
3. Clark and Hughes, *The Life and Letters of the Reverend Adam Sedgwick*, Vol. 2, p. 216.
4. Ibid., Vol. 2, p. 217.
5. Ibid., Vol. 2, p. 194.
6. The Bala group contained fossils like the trilobite *Asaphus buchii* and the brachiopod *Orthis fabellulum*. The second group contained organisms like the brachiopod *Pentamerus oblongus*, which was widely found in Murchison's Upper Silurian Wenlock formations and was sometimes known as 'Government Rock' because its broad 'arrow' shape resembled the British War Department's pre-1960s logo. I'm grateful to Colin Humphey for this information.
7. The so-called Universities Commission was set up in 1850 by an Act of Parliament to look into the statutes and revenues of the Universities of Oxford and Cambridge.
8. See *The University of Cambridge: The age of reforms (1800–82)*, Vol. 3: *Cambridge and the Isle of Ely*, London, Victoria County History, 1959, pp. 235–65.
9. Clark and Hughes, *The Life and Letters of the Reverend Adam Sedgwick*, Vol. 2, p. 220.
10. Masefield was born in nearby Ledbury in 1878.
11. Contemporary evidence suggests this was probably near Huntley Hill Quarry, a geological reserve run by Gloucestershire Geological Trust, which contains a band of what we would now call late Ordovician rock. But the contact point with the earlier Silurian rocks is not visible, and local experts are unable to confirm where Sedgwick might have found this unconformity. I'm indebted to Dave Green of Geo Studies in Gloucestershire for some of this information.
12. In 1840 John Bowman, the retired banker from Wrexham, had explored bands

of limestone on the northern slopes of the Berwyn Hills and reported that rocks assigned to the Caradoc period were indistinguishable palaeontologically from nearby Bala rocks. The following year another 'amateur' geologist, Daniel Sharpe, reached a similar conclusion on the southern slopes of the Berwyns in the valley of the River Dyfi near Dinas Mawddwy.

13. Secord, *Controversy in Victorian Geology*, p. 246.
14. Ibid., pp. 246–7.
15. Ibid., p. 246.
16. Clark and Hughes, *The Life and Letters of the Reverend Adam Sedgwick*, Vol. 2, p. 231.
17. Ibid., p. 231.
18. Secord, *Controversy in Victorian Geology*, p. 253.
19. Secord writes that Thomas Gouch of Kendal, Samuel R. Pattison of Launceston, John D. Pring of Taunton and John Giles of Liskeard all sent strong letters of support to Sedgwick. But, as Giles observed, they were just 'small folk'. Ibid., p. 254.
20. Henry De la Beche was knighted in 1848.
21. Secord, *Controversy in Victorian Geology*, p. 251.
22. Andrew Crombie Ramsey, *Proceedings of the Geological Society*, May 1904, Vol. 60, quoted in Secord, *Controversy in Victorian Geology*, p. 248.
23. William Aveline and John Salter, *Proceedings of the Geological Society*, June 1853.
24. Clark and Hughes, *The Life and Letters of the Reverend Adam Sedgwick*, Vol. 2, p. 283.
25. The names of the two strata changed repeatedly. Today we know them as the Caradoc and Ashgill.
26. Secord, *Controversy in Victorian Geology*, p. 250.
27. Geikie, *Life of Sir Roderick Murchison*, Vol. 2, p. 155.
28. Sedgwick to Murchison, October 1853, quoted in Clark and Hughes, *The Life and Letters of the Reverend Adam Sedgwick*, Vol. 2, p. 251.
29. Joseph Beete Jukes, 'Annual Address to the Geological Society of Dublin', 8 February 1854, *Journal of the Geological Society of Dublin*, Vol. VI (1853–4), pp. 61–108.
30. 'On the May Hill sandstones and the Palaeozoic System of England', Spring 1854.
31. Secord, *Controversy in Victorian Geology*, p. 267.
32. Cambridge University Library (CUL), MS Add. 7652/II/X.87.
33. Clark and Hughes write, 'we believe he did not again attend any meeting of the Geological Society'. Clark and Hughes, *The Life and Letters of the Reverend Adam Sedgwick*, Vol. 2, p. 262.

34. Ibid., Vol. 2, p. 279.
35. Secord, *Controversy in Victorian Geology*, p. 262.
36. *The Athenaeum*, October 1854, No. 1407, pp. 1243–4.
37. Secord, *Controversy in Victorian Geology*, pp. 261–8.
38. Leonard Horner to Murchison, 18 November 1855, GSL, Murchison mss.
39. Secord, *Controversy in Victorian Geology*, p. 262.
40. The classic example was his retelling of his initial finding of the Greywacke beds in the Wye Valley in 1832: see Ch. 2, pp. 42–3, in this book.
41. In 1837 he promised to send Murchison notes to support a talk he was delivering on the then-controversial issue of the rocks of Devon and Cornwall: nothing appeared, leaving an embarrassed Murchison to bluff it out. Murchison repeatedly sent Sedgwick sections of the *Silurian System* for his opinion, but got no response. By the 1850s Sedgwick had still not written a full account of the Cambrian System.
42. Sedgwick to Jukes, March 1848, *Letters and Extracts from the Addresses and Occasional Writings of J. Beete Jukes*.
43. Clark and Hughes, *The Life and Letters of the Reverend Adam Sedgwick*, Vol. 2, p. 217. See also Secord, *Controversy in Victorian Geology*, p. 264.
44. Ibid., p. 264.
45. For Sedgwick's warmer side, see his letters during the 1850s to his nieces Isabella Herschel and Fanny Hicks, and to Mary Sedgwick, the wife of his nephew, Richard. Clark and Hughes, *The Life and Letters of the Reverend Adam Sedgwick*, Vol. 2, Ch. 4.
46. Robert Burns, 'Despondency: An Ode', 1786.
47. Clark and Hughes, *The Life and Letters of the Reverend Adam Sedgwick*, Vol. 2, p. 106.
48. Letter to his friend Sir John Herschel, Winter 1853–4, in Clark and Hughes, *The Life and Letters of the Reverend Adam Sedgwick*, Vol. 2, p. 268.
49. Secord, *Controversy in Victorian Geology*, p. 221.
50. Geikie, *Life of Sir Roderick Murchison*, Vol. 2, p. 159.
51. *Dublin University Magazine*, Vol. 44 (August 1854), pp. 226–40.
52. Geikie, *Life of Sir Roderick Murchison*, Vol. 2, p. 169.
53. Quoted in James A. Secord's unpublished 'A Romance of the Field'.
54. British Geological Survey Archive, Geological Survey Material (GSM), 1/420(A).

9 The Highland Controversy: Siluria in Scotland

1. Geikie, *Life of Sir Roderick Murchison*, Vol. 2, p. 203.
2. The Highland Clearances began in the late 1700s when Scottish landowners

found they could make more money from raising Cheviot sheep than from their tenants. Tens of thousands of tenant farmers and their families were evicted from the land, often amid scenes of extreme brutality. See John Prebble, *The Highland Clearances*, Penguin, 1963.

3. Geikie, *Life of Sir Roderick Murchison*, Vol. 2, p. 650.
4. Cited in David R. Oldroyd, *The Highland Controversy*, Chicago, University of Chicago Press, 1990, p. 42. Archibald Geikie records in his biography of Murchison that Nicol was able to show 'by means of fossils that well-marked representatives of his Llandeilo and Caradoc formations occurred in the south of Scotland: a notable step in the progress of the extension of his Silurian domain over Britain'. Geikie, *Life of Sir Roderick Murchison*, Vol. 2, p. 116.
5. Metamorphism is the process during which enormous heat and pressure alter the composition of existing rocks. Minerals recrystallise and form new ones; their shape and alignment can also alter.
6. The eastern sandstone contained in particular the fossil fish Holoptychius, a marker of Devonian Old Red Sandstone.
7. Quartzite is quartz-rich sandstone transformed into a crystalline structure by metamorphism. Schists and gneiss are also classic examples of metamorphism, where intense pressure has caused the minerals to line up in a series of distinct coloured bands that are clearly visible to the naked eye.
8. 'Assynt' is said to be a Norse word meaning 'rocky ridge', a reminder that the Vikings were here around a thousand years ago.
9. Geikie, *Life of Sir Roderick Murchison*, Vol. 2, p. 204.
10. James Boswell, *The Journal of a Tour to the Hebrides with Samuel Johnson, LL.D*, first published in 1785.
11. Adam Sedgwick and Frederick McCoy, 'Synopsis of British Palaeozoic Fossils', reprinted by Cambridge University Press, 2020.
12. Geikie, *Life of Sir Roderick Murchison*, Vol. 2, p. 206.
13. Clark and Hughes, *The Life and Letters of the Reverend Adam Sedgwick*, Vol. 2, p. 304.
14. Geikie, *Life of Sir Roderick Murchison*, Vol. 2, p. 205.
15. James Nicol, *A Guide to the Geology of Scotland*, Edinburgh, 1844.
16. One of the most famous is Knockan Crag, a prominent line of cliffs running through the bogs of the Assynt, where the bands of strata were particularly well displayed.
17. See an obituary note published in *Nature*, 1879, Vol. 19, p. 590; https://www.nature.com/articles/019590a0
18. Some of the others were far less clear index fossils. They included a species of

gastropod called Euamphalus and a nautiloid mollusc called Orthoceras, neither of which was confined to the Devonian.

19. At Inchnadamph he drew a cross-section through the hills behind the hotel, which became known as the 'Cnoc an Droighinn cross-section', after the hills. At Eriboll he drew a cross-section through the hills that rise above Eriboll House on the eastern shores of the loch.
20. British Association Conference, Leeds, 1858.
21. This was particularly clear to Nicol at places like Loch Broom, and Loch Achall just east of Ullapool.
22. The red crystals would become known by the neutral and uncontroversial term of 'Logan Rock', after Loch Logan in the southern Highlands, where some were also found.
23. Henry De la Beche died in 1855 at the relatively young age of fifty-nine.
24. Geikie later remarked, beyond Murchison's hearing, that Murchison 'had demonstrated that he was not cut out for public speaking'. See Collie and Diemer, *Murchison in Moray*.
25. All quotes taken from Oldroyd, *The Highland Controversy*, pp. 87–8.
26. Murchison to Geikie, 16 January 1865, quoted ibid., p. 145.
27. Murchison to Robert Harkness, 10 January 1861, CUL, MS Add. 7652/IV/B24.
28. See the obituary note published in *Nature*, 24 April 1879, Vol. 19, p. 590; https://www.nature.com/articles/019590a0
29. James Nicol, *The Geology and Scenery of the North of Scotland*, Edinburgh, Oliver and Boyd, 1866. Nicol's appendix drew on two lectures that he gave to the Edinburgh Philosophical Institution.
30. *Nature*, 24 April 1879, Vol. 19, p. 590; https://www.nature.com/articles/019590a0
31. William Avery to Andrew Ramsey, 1856, Imperial College KGA/Ramsey/7/69; quoted in Secord, *Controversy in Victorian Geology*, p. 257.
32. Ramsey to Aveline, 1856, GSM 1/420(A).
33. Clark and Hughes, *The Life and Letters of the Reverend Adam Sedgwick*, Vol. 2, p. 285.
34. Ibid., p. 490.
35. Secord, *Controversy in Victorian Geology*, p. 270.
36. 20 January 1859.
37. Clark and Hughes, *The Life and Letters of the Reverend Adam Sedgwick*, Vol. 2, p. 373.
38. Letter to Robert Baines Armstrong, 20 May 1862, Manuscript Cumbria County Archive, Kendal, quoted in Speakman, *Adam Sedgwick – Geologist and Dalesman*, p. 83.

39. 1869, GSL: M/S, Last correspondence file.
40. Geikie, *Life of Sir Roderick Murchison*, Vol. 2, p. 314.
41. Murchison to William Whewell, 2 July 1862, Add. me.a. 213, Trinity College.
42. Clark and Hughes, *The Life and Letters of the Reverend Adam Sedgwick*, Vol. 2, p. 488.
43. Ibid., Vol. 2, pp. 441–2.
44. Geikie, *Life of Sir Roderick Murchison*, Vol. 2, p. 344.
45. Ibid., Vol. 2, p. 55.

10 The Missing Key: Discovery of the Ordovician

1. See Secord, *Controversy in Victorian Geology*, p. 307. Walter Scott lived at Abbotsford, 2 miles outside Galashields, until his death in 1832.
2. I'm grateful to Beryl Hamilton, who shared some of her research notes on Charles Lapworth with me and has written about Lapworth's friendship with the Edinburgh geologists David Page and Henry Alleyne Nicholson.
3. Cited in Oldroyd, *Earth, Water, Ice and Fire*, p. 221. See also Archibald Geikie, *Explanation of Sheet 34: The Geology of Eastern Berwickshire*, HMSO, 1863, p. 8. Geikie was appointed head of the newly established Scottish branch of the Geological Survey in 1867.
4. Archibald Geikie writes of a geologist being cut in two pieces at the mouth of a tunnel at the south end of the Longhope ridges. Geikie, *Life of Sir Roderick Murchison*, Vol. 2, p. 151.
5. Cited in Beryl Hamilton, 'Classic Paper in the History of Geology: Charles Lapworth's "The Moffat Series", 1878', *Episodes*, Vol. 24, No. 3, September 2001, pp. 194–200.
6. Charles Lapworth and James Wilson, 'On the Silurian rocks of the counties of Roxborough and Selkirk', *Geological Magazine*, 1871, Decade 1, v. 8, pp. 456–64.
7. The story comes from Lapworth's granddaughter, Patricia, in conversation with Beryl Hamilton. Details given in a lecture by Lapworth in 1879, probably to the Geological Society of Glasgow.
8. The spelling changes over time. Today it is more commonly spelt Dob's Linn.
9. From William W. Watts, *The Geological Work of Charles Lapworth*, Special Supplement to Vol. XIV of Proc. of Birmingham Nat. Hist. and Phil. Soc., 1921.
10. Charles Lapworth's 'The Silurian Age in Scotland', 2nd Lecture, p. 6, cited in Oldroyd, *Earth, Water, Ice and Fire*, p. 226.
11. Charles Lapworth, 'The Moffat Series', *Quarterly Journal of the Geological Society of London*, 1878, Vol. 34, pp. 240–346.
12. Hawkes, *A Land*, p. 49.
13. Charles Lapworth, Review of *Memoirs of the Geological Survey of the United*

Kingdom, The Silurian Rocks of Britain, Vol. 1: Scotland, 1899, in the *Geological Magazine*, Decade 4, V. 6, pp. 472–79 and 510–20, quoted in Beryl Hamilton, 'Classic Paper in the History of Geology', p. 200.

14. In fact the picture was a little more complicated than this. In order to fudge the problems posed by the May Hill sandstones, Murchison had added a third zone to the Lower Silurian in 1856, which he called the 'Lower Llandovery'. But this was a detail that didn't alter the bigger picture.
15. William W. Watts, 'Obituary Notice of Charles Lapworth', *Proceedings of the Royal Society*, B, Vol. 92, 1921.
16. Cited in Beryl Hamilton, 'Classic Paper in the History of Geology', p. 198.
17. Charles Lapworth, 'On the tripartite classification of the Lower Palaeozoic rocks', 1879, quoted in Watts, *The Geological Work of Charles Lapworth*.
18. It has been suggested there may have been an extremely severe Ice Age in the southern hemisphere, which caused a fall in sea levels, a change in ocean currents and a corrresponding change in the climate.
19. While Archibald Geikie remained director, the British Geological Survey refused to acknowledge the legitimacy of Lapworth's Ordovician System. It was only when he was replaced by J. J. H. Teall in 1902 that the Survey recognised Lapworth's new period. The International Union of Geological Sciences (IUGS), an international NGO, was founded in 1961 and is committed to international cooperation in the field of geology. Currently 121 countries or regions of the world are represented.
20. Quote taken from an anonymous letter, signed simply 'A Field Naturalist', to the *Daily Chronicle*, 6 February 1893.
21. Hawkes, *A Land*, p. 54.
22. Oldroyd, *Earth, Water, Ice and Fire*, p. 240.
23. Charles Lapworth, 'The Secrets of the Highlands', *The Geological Magazine*, Vol. 10, Issue 8, August 1883, p. 125.
24. These categories further subdivided into two more: a discrete band of orange silts and muds containing the impressions of what were once thought to be fucoids – a primitive form of bladderwrack, now believed to be the trails of long-dead worms – known today as the Fucoid Beds; and a second band of gritty sandstone containing the remains of an early 'snail' called *Salterella maccullochi*, which became known as the Salterella Grits.
25. Lapworth, 'The Secrets of the Highlands', p. 124.
26. Today the best place to view this cross-section through the rocks of the region in much greater detail is further west at Knockan Crag in the Assynt. Here eight bands of strata that make up the rocks of northern Scotland are clearly visible.
27. Letter from Lapworth to his friend Thomas Bonney, September 1882. Bonney

was another former teacher who had risen from the status of 'amateur geologist' to become professor of geology at University College London and a leading member of the Geological Society.

28. Some of the best exposures of Mylonite are at places like Sangobeg Bay on the coast road between Durness and Eriboll where, at the foot of a steep sandbank rising up from the beach, there are outcrops of dark rock with highly visible fine, pale threads running through them like a series of lamina.
29. Delivered in March 1883.
30. All quotes come from Sir Jethro Teall, cited in Watts, 'Obituary Notice of Charles Lapworth'.
31. Watts, *The Geological Work of Charles Lapworth*, p. 41.
32. Letter from John Judd to Lapworth, 20 September 1882, Lapworth Archive, Catalogue no. A4.
33. Sir Jethro Teall, quoted in Watts, 'Obituary Notice of Charles Lapworth'.
34. See Oldroyd, *The Highland Controversy*, pp. 250–51, for a discussion of this.
35. Watts, *The Geological Work of Charles Lapworth*, p. 43.
36. Archibald Geikie, *Textbook of Geology*, first published in 1880.
37. Watts, *The Geological Work of Charles Lapworth*, p. 24.
38. The point where the two plates finally join is known as the Iapetus Suture. It runs slightly to the south of the Moine Thrust, from the Solway Firth in the west to St Abb's Head on the east coast of Scotland, just south of Edinburgh and close to James Hutton's famous unconformity at Siccar Point.
39. Watts, *The Geological Work of Charles Lapworth*, p. 41.

11 The Final Mystery

1. There is some confusion about the exact number of cables. Most reports talk of twelve cables, but Bruce Heezen at Columbia University, who investigated the earthquake twenty-three years later, talked of thirteen cables.
2. The first man to speculate about the existence of underwater avalanches was a Canadian geologist called Reginald Daly, working at Harvard University in the early 1920s.
3. B. C. Heezen and M. Ewing, 'Turbidity currents and submarine slumps, and the 1929 Grand Banks [Newfoundland] earthquake', *American Journal of Science*, 1952, Vol. 250, p. 849.
4. Such deep-water pebbly, sandy rocks are now called 'Turbidites'.
5. See, for instance, Peter Toghill, *The Geology of Britain*, Shrewsbury, Airlife Publishing, 2002, p. 16.
6. *Edinburgh Review*, 27 May 1867, CUL, MS Add. 7652/V/f.1.
7. Ibid.

8. Rudwick, *The New Science of Geology*. In Varium Collected Studies Series, Farnham, Ashgate, Ch. XI, p. 243.

LIST OF ILLUSTRATIONS

ACKNOWLEDGEMENTS

It would have been impossible to write this book without four inspirational and detailed accounts of the subject: James Secord's *Controversy in Victorian Geology: The Cambrian–Silurian Dispute* (Princeton University Press, 1986); Martin Rudwick's *The Great Devonian Controversy* (University of Chicago Press, 1985); Michael Collie and John Diemer's *Murchison's Wanderings in Russia* (British Geological Survey, 2004); and David Oldroyd's *The Highland Controversy* (University of Chicago Press, 1990). I have benefited enormously from all of them. In addition, Jim Secord has gone out of his way to be supportive and helpful, lending an enormous box of his research notes, providing unpublished material, and reading and commenting extensively on many of the chapters. Martin Rudwick and John Diemer have read chapters and made helpful suggestions. So too has Beryl Hamilton, who has written extensively on Charles Lapworth and also shared some of her research notes with me.

I am not an expert on the rocks, but practically every geologist I have approached has been generously willing to share their time and knowledge. Hugh Torrens, a fount of wisdom on (among other things) the history of the exploration of the rocks of the English–Welsh border region, gave advice and encouragement and, very importantly, introduced me to the geologist Geoff Townson, who spent many patient hours educating, commenting and advising me on the sedimentary rocks of England. I am extremely grateful to him. That some of this material didn't make it into the book is no reflection on the quality of his advice.

I have had a long and fruitful correspondence with Colin Humphrey, formerly associated with the Mid Wales Geology Club, who offered help and advice. Charlie Bendall, of the same club, provided information on the Bala Limestone. Mike Rosenbaum and Robert Williams of the Shropshire Geological Society talked to me about the exploration of the geology of the region. Duncan Hawley took time to send a detailed note on the Greywacke–Old Red Sandstone boundary in southern Wales and on Murchison's apparent discovery of a similar boundary in the Wye Valley.

Jon Clatworthy, director of the Lapworth Museum of Geology in Birmingham, guided me through their enormous Lapworth archive; Caroline Lam, archivist at the Geological Society, opened their comprehensive Murchison collection; and Sandra Freshney, archivist at the Sedgwick Museum in Cambridge, dug out Sedgwick's carefully preserved notebooks. A huge thanks to you all.

I would also like to thank Daniel Lockett of the Ludlow Museum Resources Service for information about the geology of the region and Murchison's trips to the town and its surrounding rocks; Kevin Page of Plymouth University, who helped locate some of the sites probably visited by Sedgwick and Murchison in Devon and Cornwall; and Dave Green of Geo Studies in Gloucestershire for insights into the geology of May Hill.

When I put out a notice on the History of Geology website, I received a totally unexpected email from Arthur Murchison in the US, a possible distant relative of Roderick Murchison and an expert on his military career. It generated an email exchange, which opened my eyes to the military culture that pervaded Murchison's worldview and, unexpectedly, to Sedgwick's rather active role as a senior proctor and upholder of morals in Cambridge.

Andrey Kurochkin has done a fabulous job with the maps and diagrams. Kathy Henderson and Al Kerr read various versions of the text, pointed out stupid mistakes and suggested improvements. So have Jonathan Wilkins and Douglas Palmer. And Mandy Greenfield has cleaned up the text wonderfully. I am very grateful to all of them. Finally a conversation with Ruthie Petrie nudged me towards Profile Books, and Ed Lake, my editor at Profile, has made it a better book than it would have been without his help and advice.

INDEX

Page references for illustrations are in *italic*; those for notes are followed by n

Index

Index